REMOTE RESEARCH

Nate Bolt

Tony Tulathimutte

Rosenfeld Media
Brooklyn, New York

Remote Research
By Nate Bolt and Tony Tulathimutte

Rosenfeld Media, LLC
457 Third Street, #4R
Brooklyn, New York
11215 USA

On the Web: www.rosenfeldmedia.com
Please send errors to: errata@rosenfeldmedia.com

Publisher: Louis Rosenfeld
Editor: Marta Justak
Copy Editor: Chuck Hutchinson
Interior Layout Tech: Danielle Foster
Cover Design: The Heads of State
Indexer: Nancy Guenther
Proofreader: Dan Foster

DEDICATIONS

Nate: To my family, friends, colleagues, teachers, and loved ones

Thanks for listening to me ramble on about this field for so long.

Tony: To Mike Hardnett and Renee Zalles

HOW TO USE THIS BOOK

Who Should Read This Book?

This book is about *remote user research*, which is a method of using Internet tools and services to conduct user research with participants who are in another location. (User research, in turn, is the field of studying how people interact with technology.)

Are you a user experience/human-computer interaction practitioner? If so, you're totally gonna love this book, especially if you've ever been frustrated with current in-person or lab methods of user research for any of the several reasons we describe in Chapter 1. If you're a software or Web developer looking for insights into your own (or your competitors') designs, or an interaction designer or consultant, you'll probably dig this book too.

Is there anyone this book *isn't* for? You don't *need* to be a veteran user-experience researcher to understand what we talk about in this book, although we do focus mostly on the "remote" aspects of remote research. You won't find much advice on how to conduct user research *in general*—for that, a great place to start is Mike Kuniavsky's *Observing the User Experience*.

What's in This Book?

Remote Research is a how-to book about remote research methods: using a phone and the Internet to conduct user experience research from a distance.

In the **Introduction and Chapter 1**, you'll get an overview of what remote research is all about, when you should and shouldn't use remote methods, and the two main kinds of remote research studies: *moderated* and *automated*.

In **Chapters 2 through 5**, you'll learn how to set up, recruit, and conduct a basic remote moderated study. We describe a method called "live recruiting," which involves intercepting visitors to your own Web site to participate in your studies immediately. We also discuss the privacy and consent issues around recruiting, session recording, and remote participation.

In **Chapter 6** we describe various automated research methods, illustrating them with case studies.

Chapter 7 provides advice about how to analyze and report on the findings of remote research studies.

Chapter 8 is a short guide to tools and services you can use to fill the many technological needs of remote research, including screen sharing applications, recording software, and several online automated research Web apps.

Chapter 9 illustrates how many of the basic principles described earlier in the book can be adapted and applied to special testing circumstances, where normal remote testing methods aren't possible or desirable.

Chapter 10 is a review of the major challenges you'll face when planning, conducting, and presenting a remote research study.

What Comes with This Book?

This book comes with two companion Web sites. The first is the Remote Research page (🄰 www.rosenfeldmedia.com/books/remote-research) on the Web site of our publisher, Rosenfeld Media. The other site is our companion Web site for the book, Remote Usability (http://remoteusability. com). On both sites you'll be able to find detailed resources, research document templates, supplementary articles, hi-res figures and diagrams, and up-to-date lists of remote research tools and services.

FREQUENTLY
ASKED QUESTIONS

What is remote user research, anyway? Is it anything like focus groups or surveys?

Remote user research is simply a type of user experience (UX) research that's conducted over the phone and Internet, instead of in person. In general, UX research seeks to understand how people interact with technology. Unlike focus groups and surveys, market research techniques that are used to learn people's opinions and preferences, UX research focuses on studying people's behavior. In that sense, remote user research isn't really like market research; however, both remote user research and market research can be applied toward improving the design of existing technologies and inspiring new ones. See Chapter 1, page 3.

What kinds of remote research are there?

That's a huge question, and we spend a good chunk of this book introducing and describing the many varieties and specialties of remote research out there. In general, there are two branches of remote research: moderated and automated. In a moderated study, a researcher talks directly to the participants as they use the interface in question, and it's good for obtaining rich, qualitative feedback. In an automated study, you use online tools and services to gather behavioral or written feedback and information automatically, without the researcher's direct involvement. For more about moderated testing, see Chapter 2, page 32, and all of Chapter 5. For more about automated testing, see Chapter 6.

I'm skeptical about remote research. If it's so great, why haven't I heard of it?

A lot of the misgivings that people have about remote research come from its novelty. The method is still cutting edge, and the technique requires a certain degree of know-how. Until now, there hasn't been a book you could learn the method from—which is, of course, the reason we wrote it.

Still, lots of companies have done it, with great success. We've done remote studies with Sony, Autodesk, Greenpeace, AAA, HP, Genentech, Wikipedia, UCSF Medical Center, the Washington Post, Esurance, Princess Cruises, Hallmark, Oracle, and Blue Shield of California, among many others.

I'm still skeptical. Can you really get valid behavioral feedback without seeing your participants in person?

Since remote research is conducted over the phone and Internet, many people worry about missing "rich details" like facial expressions and body language. First, we believe that for most user research studies, the way that users interact with the interface and their think-aloud comments are the only really necessary things to focus on. And on top of that, much of the tone does come through the user's voice and language. We weigh the pros and cons of in-person research and remote research in Chapter 1, pages 5-15, and discuss moderating over the phone in Chapter 5, page 110.

Still skeptical? Then you should check out our exhaustively documented study for Wikipedia, complete with full-session videos and highlight clips at http://usability.wikimedia.org/wiki/Usability_and_Experience_Study. It includes both lab and remote sessions with identical goals, so it's a good comparative case study.

The best way to see if a remote study is for you, however, is by getting your feet wet with a quick, painless pilot study, which we'll walk you through in Chapter 2, page 30.

I want to cut costs for my user research study. Is using remote methods a good way to do that?

Not really. You might save on costs related to travel expenses, renting a lab facility, or hiring a recruiting agency, but then there are the expenses of the specialized remote research tools and services you'll need, and where the researcher's time, participant incentives, and project timeline are concerned, nothing is much different. We cover the equipment requirements of a basic moderated study in Chapter 2, page 28, and the costs of many remote tools and services in Chapter 8.

Where can I get people to participate in my remote study?

You can technically use any methods to recruit for a remote study that you'd use for an in-person study: email contacts, recruiting agencies, and even craigslist ads (blegh!) are still an option. However, in this book, we introduce a method called *live recruiting*, with which you intercept visitors to your Web site by using a pop-up form to get them to participate in your study right away. We strongly believe this approach is ideal for remote research because it allows you to do what we call "Time-Aware Research," which we introduce in Chapter 1, page 10 and discuss in depth in Chapter 3.

How on earth can you call this a book about research without way more academic references and doctoral degrees?

This book is aimed at people who want to hear practical information about how remote research is done in the real world.

We don't claim absolute peer-reviewed scientific rigor, but we are sharing what's given us the best results after almost a decade of remote research experience: 2,615 moderated users, 2,676 automated users, 234 projects, 19,120 project hours, and 89 clients to date. Simply put, this is just what works for us, and we think it will work for you, too. (And for what it's worth, Nate and Tony both have cognitive sciences/human-computer interaction degrees: yes, we roll that deep.)

TABLE OF CONTENTS

CHAPTER 4
Privacy and Consent   **73**

CHAPTER 5
Moderating   **93**

CHAPTER 10
The Challenges of Remote Testing   237

CONCLUSION
Don't Waste Your Life Doing
Pointless Research   249

FOREWORD

Peter Merholz is a founding partner and president of Adaptive Path, an experience strategy and design firm (www.adaptivepath.com). *He coauthored* Subject To Change: Creating Great Products and Services for an Uncertain World, *published by O'Reilly.*

You are holding in your hand a portal into the future. The approaches discussed in this book are the first steps toward a user research methodology suited to our technological context. In comparison, lab research will be increasingly seen as an archaic approach to understanding people, akin to phrenology or trepanning.

You see, lab usability engineering was born of a simpler time. Files were stored on floppy disks, and the computer wasn't connected to a network. People used computers for Calculation (basic math, spreadsheets, etc.), Creation (word processors, graphics programs), and Capture (data entry). Computers would have only one "program" running at a time. And someone could be expected to focus on a single task at hand for many minutes, if not hours on end.

These days, our customers' technological world is much more complex. The bulk of their time online is spent engaged in Consumption (browsing the Web, listening to music) and Communication (email, instant messaging, Twitter), though they still Calculate, Create, and Capture. They have multiple applications open and multiple windows within those applications. Thousands of files and emails fill their hard drives, and they're managing multiple devices, including computers, mobile phones, iPods, and digital cameras. Technologies have driven users to a point of extreme distraction—recent research has shown that workers are interrupted an average of every 11 minutes.

Although the world has changed, the methods of standard usability practice are essentially the same as were practiced in the early 1990s. We recruit participants who satisfy demographic requirements, invite them into a fluorescent-lit lab with an imposing mirror along one wall, ask them to use a computer they've never seen before, have them engage in a set of scripted tasks, and "think aloud" while doing so. These participants won't have an IM window pop open unannounced, nor be able to click on the bookmark bar to see their Facebook page, nor be able to pull up the notes they saved in a draft in their email. When everything about the observation environment is so unnatural, how can we expect our findings to provide legitimate insights?

This is where the genius of Nate Bolt and his crew come in. Since I've known them, they've experimented with technologies in an effort to capture the most authentic user research data. Whether it was the pioneering screen sharing and recording tool Ethnio (which fundamentally changed how I considered usability), or real-time uploading of video captured in a drive-along study for Volkswagen, their mission is to deliver the unvarnished truth.

Now, it's easy to get caught up in the tools. While remote research is enabled by nifty technologies, which should be of secondary interest at best, I know that for Nate and Tony, this research approach is about mindset. When you simply desire to understand people in the most direct ways, you come up with clever means of doing so. And sometimes, you can't go "remote." When I first met Nate, he crowed about how they kitted out a usability lab with hidden microphones and cameras so that the participant would be as comfortable as possible. Or when researching the use of multiplayer video games, he recreated a cozy living room set up in his office, with snacks and lounge chairs, to make gamers feel at home.

And that's how this book is a time machine. Only the Flying Spaghetti Monster knows what the future will bring. The only thing that's certain is that it's going to keep changing. To prepare for that future, you need the mindset represented in this book, in order to figure out which approaches will provide the best user research data given your circumstances. Like Bolt | Peters, don't settle for standard practices; instead, play with new ideas that will deliver fresh insights.

—Peter Merholz

CHAPTER 1

Why Remote Research?

I n-person lab research used to be the only game in town, and as with most industry practices, its procedures were developed, refined, and standardized, and then became entrenched in the corporate R&D product development cycle. Practically everything gets tested in a lab nowadays: commercial Web sites, professional and consumer software, even video games (see Figure 1.1).

FIGURE 1.1
Brighton University's usability lab, from behind the traditional two-way mirror.

The Appeal of Lab Research

Part of the appeal of lab-based user research was that it provided a seemingly scientific basis for making decisions by using observational data, instead of someone's error-prone gut instincts. Stakeholders appreciated the firm protocol and apparent reliability of properly managed lab research. Lots of user research practitioners continue to perform lab research just because it's what people have been doing for a long time.

Market Research vs. User Experience Research

Let's make something clear. Focus groups are practically synonymous with user research in most people's minds, and focus groups belong to the world of market research. But there's a huge difference between market research and user experience (UX) research. Market research is much more common and comprises the lion's share of research spending; UX research comprises just a fraction of that (see Figure 1.2).

THE WORLD *of* RESEARCHING PEOPLE

Market Research	UX Research
Opinions	**Behavior**
* Focus groups	* Ethnography
* Surveys	* Think-aloud tasks
* Preference interviews	* Conceptual (e.g., card sorting)
* Ad/brand awareness	
* Concept testing	* Concrete (e.g., usability on a live Web site)
* Ideation	* Task elicitation

FIGURE 1.2
The relationship between market research and UX research. The two fields seem similar, but they have different goals and take different forms.

However, *this book is about user experience research, not market research.* The main difference between the two fields is that market research focuses on opinions and preferences, whereas UX research focuses on behaviors. The distinction can be confusing, especially since a lot of online consumer research companies try to convince you they give you insight into "what your customers are doing on your Web site," when they're really just providing opinions.

Market Research vs. User Experience Research (continued)

A market research study might have goals like these:

- "Determine how users respond to our branding."
- "Identify different segments' color preferences for the homepage."
- "See if users like our new mascot."
- "Determine what users enjoy most and least about our site."

While the goals of a UX study, on the other hand, would sound more like these:

- "Can anyone actually use my interface?"
- "Determine where users make errors in completing a purchase."
- "See whether users can successfully create a playlist."
- "Understand why users aren't logging in."
- "See how users mentally organize different product categories."

It's important to keep in mind that market research is pretty useless over small sample sizes. Opinions can vary widely across demographics and location, are very sensitive to the phrasing of the research questions, and can change fairly quickly. Behavior, on the other hand, is fairly consistent across demographics and location for many tasks, and most usability flaws in a given software interface can be uncovered in a moderated study by a much smaller number of users.

How small? That's a contentious question. UX luminary Jakob Nielsen (in)famously claimed that having five users was enough to uncover 80% of usability flaws in an interface, but others like Jared Spool insist that the number depends on factors such as user segmentation, risks associated with the errors, task complexity, and so on. At any rate, our point is that a moderated UX study usually requires much fewer people than a market research study.

Market Research vs. User Experience Research (continued)

Put it this way: ask 10 people what they think about how well a door is designed, and their comments might not overlap at all. One blames the condition of the hinges, another talks about the weight of the door, another complains about the color of the doorframe, and so on. But if you observe 10 people walking through the door, and the first two accidentally try to push when they ought to pull, then you've found your design flaw right there.

So, to put it all together: whether you go with market research or UX research depends on what you're trying to find out. This book is about UX research, so it's focused on user behaviors rather than opinions.

FURTHER READING ABOUT THE SAMPLE SIZE QUESTION

Turner, C. W., Lewis, J. R., and Nielsen, J. (2006). Determining usability test sample size. In W. Karwowski (ed.), *International Encyclopedia of Ergonomics and Human Factors* (pp. 3084–3088). Boca Raton, FL: CRC Press.

Lewis, J. (2001). Evaluation of procedures for adjusting problem-discovery rates estimated from small samples. *The International Journal of Human–Computer Interaction* 13(4), 445–479.

Lindgaard, G., and Chattratichart, J. (2007). Usability testing: What have we overlooked? In *Proceedings of the SIGCHI Conference on Human Factors in Computing Systems* (San Jose, CA, USA, April 28–May 03, 2007). CHI '07. ACM, New York, NY, 1415–1424.

Is Lab Research Dead?

Heck no. Lab and remote research share the same broad purpose: to understand how people interact and behave with the thing you've made (from here on, let's just call it "the interface"). There's no need to set up a false opposition between the two approaches—one isn't inherently better than the other. Despite the versatility of remote research, there are lots of reasons you might want to conduct an in-person study instead, most of which have to do with security, equipment, or the type of interaction you want to have with your research participants. More generally, lab research is appropriate when you need a high degree of control over some aspect of the session, such as the following situations.

Info security. Security is often a concern for institutions like banks and hospitals, which deal in sensitive information, or companies concerned with guarding certain types of intellectual property. If you're testing a top-secret prototype, you obviously don't want to let people access something from their home computer, where it could be saved or screen-captured. On the other hand, you might also be doing a study on users who would be secretive about sharing what's on their screen—government employees, doctors, or lab technicians, for instance. Either way, you'll want to test users in a controlled lab environment to keep things confidential, especially if what you're testing is so hush-hush that you must have your users sign a nondisclosure form.

Inability to use screen sharing. You might also want to use a lab if your users are unable to share their screen over the Internet, for whatever reason. Some studies (of rural users, cybercafe patrons, etc.) may require you to talk to users who don't have reliable high-speed Internet connections, who own computers too slow or unstable to use screen sharing services effectively, or who have operating systems incompatible with the screen sharing tools you're using. These restrictions apply only to moderated studies, for which you need to see what's on your users' screens.

The need for special equipment. Depending on the interface you're testing, you may require certain special software or physical equipment to run the study properly, which is most often the case with software that's still under development. Getting users to install and configure tools to run elaborate software can be a pain (though that's not unheard of), and requiring users to have certain equipment can make recruiting needlessly difficult.

The importance of seeing the user's body. Some kinds of research will require you to study certain things about the user that are difficult to gather remotely. UX research has recently begun using eye-tracking studies, and for that kind of study, you'd need to bring the users to the eye-tracking device. Other studies might require you to attend to the participants' physical movements, which may be difficult to capture with a stationary webcam. And then there are multiuser testing sessions, in which a single research moderator facilitates many participants at once; screen sharing is currently not well suited to sharing multiple desktops at once, though some tools (e.g., GoToMeeting) make it relatively painless to switch from one desktop to another. We want to emphasize, however, that for most studies, seeing the user at all is *not* actually important; we explain why in Chapters 5, "Moderating," (see "Ain't Nothing Wrong with Using the Phone") and 10, "The Challenges of Remote Testing."

Although these situations are all compelling reasons to conduct in-person research, part of what we want to demonstrate in this book is that remote research is very broad and adaptable, and even if a study is conducted in a lab, elements of remote methods can be adapted and incorporated to enhance in-person research methods. We'll get to that in Chapter 9, "New Approaches to User Research."

Why I Went Remote

by Brian Beaver

Old habits die hard, but for any number of reasons—cost, convenience, international testing—a handful of former lab researchers have switched to remote methods and never looked back. **Brian Beaver**, *award-winning creative director at Sony, explains why he decided to go with remote methods after many years of lab testing.*

ON GOING REMOTE

I have quite a bit of experience, either organizing user research sessions or participating in them, especially at Sony. My research has been pretty varied and the outcomes are always interesting, but I'm a big fan of remote usability testing. It seems to give me the best bang for my buck.

I'd read about it a few years back and had done some work with Adaptive Path when it had a focus on usability. Adaptive Path referred me to Nate [Bolt, coauthor of this book], and when he shared his remote approach with me, I knew we were in sync because a lot of the pain points and skeptical raised eyebrows around results we'd obtained in previous lab testing instantly diminished with remote usability testing.

Why I Went Remote (continued)

The pain points always involved recruiting. With a Web site you have such a diverse geographic base that it can be challenging to bring a core group of your users together in one location. Sony tends to be very protective of its customer information, and wouldn't share it with a research company for the purposes of recruiting, so we'd have to take on that task ourselves, which was always painful.

The raised eyebrows were always about participant motivation and validity of the recruiting process and methodology. There were always questions: How valid are these findings? Are these real users? But when you're intercepting users who are on your Web site in the middle of performing a task, those questions evaporate.

ON PARTICIPATING IN REMOTE RESEARCH

In the past we'd invite our business partners or stakeholders to the lab, but it was difficult to get them to take time out of their day to travel to the lab, and it was a big production. But if they can just bring their laptop to a conference room down the hall and just be there to listen in, it's fantastic. You'd get the same advantages if you had everyone available to go to the lab testing, and the level of engagement is a lot greater. By having a lot of stakeholders in the room, you get more diverse viewpoints, and the interaction between us observers and the moderator tends to be lively—we chat throughout that whole interview. The ability to observe and discuss things as they come up and then immediately give feedback to the moderator is really powerful.

Because we are involved in the research process, we've got our customers and the usability to consider on the one hand, and on the other hand we have a lot of business stakeholders who have strong opinions about how things should be done. So having everyone in the room watching the feedback and engaging with the process is really powerful. We were recently in the middle of a digital camera usability session and were asking the user to go through the features and content we have on the site, and the customer's going through it and he's like, "This all seems really impressive, but I really just want to know if it takes great pictures." And you see this light bulb go off above the product marketing people's heads. We're so close to this that we have absolute myopia. It was a real eye-opening moment.

Why I Went Remote (continued)

ON BENEFITING FROM REMOTE METHODS

One study was about TVs and the TV shopping process. Sony has a broad line of TVs, somewhere around 9 to 10 different series, and each has a dozen size options, so you have a lot of choices. During the study there was an "Ah-ha!" moment, a phenomenon we haven't seen before: people would often have half a dozen to a dozen sites already open when we contacted them, and they were seamlessly going between sites like Engadget, CNET, Sony, Circuit City, Best Buy, really taking advantage of browser tabs to cross-shop and gather information. We simply wouldn't have gotten that insight from a lab environment because we wouldn't have been intercepting people in their natural browsing environment; instead, they'd sit down, have the browser already open, and they'd go. So that behavior would have been completely missed.

The outcome was that, knowing that customers are looking not only for customer reviews but trustworthy, third-party editorial content, we're actively pursuing ways to bring that content into the SonyStyle site, so that from within the interface they can access that info, instead of relying on the multi-tab approach. In the past, if a product was awarded an editor's choice, we would have put that on the page as a badge of honor, but I doubt that we would have ever actually included the editorial alongside the product, if it hadn't been for this study.

ADVICE FOR THOSE CONSIDERING GOING REMOTE

If we're talking about remote testing for Web sites, from my perspective it's really a nonchoice. Having the benefit of intercepting users that are already coming to your site in order to perform a task already puts you so far ahead of the game because the motivation is there, you've got them captive, and you just gain so many more insights compared to creating an artificial environment with artificial motives. So you know from the quality and granularity of the results you're going to get, it's so much richer. If given the choice, I'll never go back to lab testing again. And there's the cost savings. Clearly, overall, it's a less costly proposition. You avoid all the travel costs. There's always a dud user in every batch of lab participants, and the great thing with usability testing is, if you start talking to someone you want to cut loose, it's no harm; you can move on to the next person, as the recruiting form is literally filling up before your eyes.

What's Remote Research Good For?

Again, most studies can successfully be done *either* in person or remotely, but, just as there are times when lab testing is more appropriate, there are also times when it makes more sense to use remote research methods.

Time-Aware Research

Remote research is more appropriate when you want to watch people performing real tasks, rather than tasks you assign to them. The soul of remote research is that it lets you conduct what we call *Time-Aware Research (TAR)*. By now, UX researchers are familiar with the importance of understanding the usage context of an interface—the physical environment where people are normally using an interface. Remote research opens the door to conducting research that also happens at the *moment* in people's real lives when they're performing a task of interest. This is possible because of live recruiting (the subject of Chapter 3), a method that allows you to instantly recruit people who are right in the middle of performing the task you're interested in, using anything from the Web to text messages. Time-awareness in research makes all the difference in user motivation: it means that users are personally invested in what they're doing because they're doing it for their own reasons, not because you're directing them to; they would have done it whether or not they were in your study.

Consider the difference between these two scenarios:

- You've been recruited for some sort of computer study. The moderator shows you this online map Web app you've never heard of and asks you to use it to find some random place you've never heard of. This task is a little tricky, but since you're sitting in this quiet lab and focusing—and you can't collect your incentive check and leave until you finish—you figure it out eventually. Not so bad.

- You've been planning a family vacation for months, but you've been busy at work so you procrastinated a bit on the planning, and now it's the morning of the trip and you're trying to quickly print out directions between finishing your packing and getting your kids packed. Your coworker told you about this MapTool Web site you've never used

Time-Aware Research.

Recruit someone who's
in the middle of a task.

Observe their behavior.

before, so you decide to give it a shot, and it's not so bad—that is, until you get stuck because you can't find the freaking button to print out the directions, and you're supposed to leave in an hour, but you can't until you print these damn directions, but your kids are jumping up and down on their suitcases and asking you where everything is. Why can't they just make this stupid crap *easy to use?* Isn't it *obvious* what's wrong with it? Haven't they ever seen a *real person* use it before?

Circumstances matter a lot in user research, and someone who's using an interface in real life, for real purposes, is going to behave a lot differently—and give more accurate feedback—than someone who's just being told to accomplish some little task to be able to collect an incentive check. Time-awareness is an important concept, so we'll bring it up again throughout this book to demonstrate how the concept relates to different aspects of the remote research process (recruiting, moderating, and so on).

> **NOTE** TAR KEEPS YOU IN THE RIGHT 1985
>
> Remember that diagram in *Back to The Future II*? Doc argues that messing with time has sent the world crashing hopelessly toward an alternate reality where things are horrible—the "wrong 1985." And that's sort of what happens when you try to assign people a hypothetical task to do at a time when they may or may not actually want to do it: you're meddling with their time, and it'll create results that look like the real thing but are all wrong.
>
> When you schedule participants in advance and then ask them to pretend to care, you're sending your research into the wrong 1985. If you don't want to create a time paradox—thereby ending the universe—you should do time-aware research.

Other Benefits of Remote Research

Here are some additional benefits of remote research.

Geographic diversity. Even if you do have a lab, the users you want to talk to may not be able to get to it. This is actually the most common scenario: your interface, like most, is designed to be accessed and used all around the world, and you want to talk to users from around the world to get a range of perspectives. Will Chinese players like my video game? Is my online map widget intuitive even for users outside Silicon Valley? Big companies like Nokia and Microsoft are often able to conduct huge, ambitious research projects to address these questions, coordinating research projects in different labs around the world, flying researchers around in first-class. If you don't have the cash for an international Gorillas-in-the-Mist project, then remote research is a no-brainer solution. If you can't get to where your users are, test them remotely.

Ability to test almost anywhere. Remote research has comparatively minimal setup requirements and can reach anywhere that computers and the Internet can go: you can be anywhere; your participants can be anywhere. Lone-wolf consultants and start-up teams working out of cafés can have trouble finding the quiet office space they need to do in-person testing. If it's too much bother to set up a proper lab, go remote; all you'll need is a desk.

Some reduced costs. Beyond travel expenses, other costs associated with lab testing may be reduced or eliminated when you test remotely. With live recruiting methods, you can get around third-party recruiting costs, and because the recruiting pool is larger, you may not have to offer as much in the way of incentives as you might otherwise to attract enough participants. Because sessions are conducted through the computer, you can use relatively inexpensive software to replace costly testing accessories, such as video cameras, observation monitors, and screen recording devices. (Note, however, that the overall cost of a remote research study is often comparable to an in-person study for many reasons; see Chapter 10 for reasons why.)

Quicker setup. Closely related to the issue of money, as always, is time. Nearly all existing recruiting methods take many weeks. Recruiting agencies usually require a couple of weeks to gather recruits, and writing out precise recruiting requirements and explaining the study to them can eat up a lot of time. Getting users from your own mailing list can be faster and moderately effective, but what if you don't have one? Or what if you've overfished the list from previous studies, or you don't want to spam your customers, or you're looking to test people who've never used your interface or heard of your company before? In any of these cases, recruiting your users online makes a lot of sense, since it allows you to do your recruiting as research sessions are ongoing. (We teach you how to do all this in Chapter 3.)

Context-dependent interfaces. Some interfaces just don't make any sense to test outside their intended usage environment. If you need users to have their own photos and videos to use in a video editing tool, having them bring their laptop or media to a lab will be a tremendous hassle. Or, let's say you're testing a recipe Web site that guides users step-by-step through preparing a meal; it wouldn't make much sense to take people out of their kitchen, where they're unable to perform the task of interest. When this is the case, remote research is usually the most practical solution, unless the users *also* lack the necessary equipment.

When to Go Remote

If you have the gumption, you can test almost anything remotely. There are ways to get around nearly any obstacle, but the approach you take is all about what's most practical and accurate. If it's significantly cheaper, faster, or less of a hassle for you to just bring people into a lab, then by all means bring 'em in. Sometimes this decision can be a tough call; users in the developing world may have limited access to the Internet, for instance, so you'd have to decide whether it's worthwhile to fly over and talk to users in person, or to find people from that demographic in your area, or to arrange for the users to be at a workable Internet kiosk to test them remotely.

For clarity's sake, let's talk about some clear-cut cases of things you should and shouldn't test remotely.

Remote testing is a no-brainer for Web sites, software, or anything that runs on a desktop computer—this is the kind of stuff remote research was practically invented to test. The only hitch is that the participants need to be able to use their own computer to access whatever's being tested. Other Web sites besides your own are a cinch: just tell your users during the session to point their Web browsers to any address you want. If you're testing prototype software, there needs to be a secure way to digitally deliver it to them; if it's a prototype Web site, give them temporary password-protected access. If the testing is just too confidential to give them direct access on their computer, you can host the prototype on your own computer and use remote access software like VNC or GoToMeeting to let them have control over the computer. There's almost always a way to do it.

The stuff you test doesn't even have to be strictly functional. Wireframes, design comps, and static images are all doable; we've even tested drawings on napkins (really). Just scan them in to a standard image format and put them on a Web site. Make sure the user's browser doesn't automatically resize it by using a plain HTML wrapper around each image. There are also plenty of software solutions (like Axure and Fireworks) that can help you convert your images to HTML.

Can you test programs that require users to enter personal information? Yes, but make sure to give your participants a way to enter "dummy" information wherever they're required to enter sensitive or personally identifying information. (According to Rolf Molich, people act differently when using dummy information, so bear that in mind.) If you require the participant to use real personal information, be sure to obtain explicit consent right at the beginning of the testing session (an issue covered in Chapter 4); you don't want to spend 20 minutes on the phone with a user only to have to terminate the study over privacy issues.

Most remote research tools (screen sharing, recording, chat, etc.) are suited for a computer desktop environment, so physical products are harder to test remotely. We're just beginning to see mobile device and mobile interface research become feasible, and we've researched interfaces like cars and computer games using some remote research methods. Plus, webcams, Web video streaming, and wireless broadband are all becoming more accessible, so there's plenty of hope. But physical interfaces will require you to come up with some creative solutions and workarounds beyond the standard remote desktop testing approach. These approaches may or may not be worthwhile; see Chapter 9 for examples of some of these alternative remote approaches.

Case Study: Lab vs. Remote

By Julia Houck-Whitaker, Adaptive Path (and Bolt | Peters alum)

In 2002, Bolt | Peters conducted two remote studies on the corporate Web site of a Fortune 1000 software company. Both studies used identical test plans, but one was executed in a traditional usability lab, whereas the other was conducted remotely using an online screen sharing tool.

SUMMARY

Our comparison showed key differences in the areas of time, recruiting, and resource requirements, as well as the ability to test geographically distributed user audiences. Table 1.1 summarizes the key differences we found comparing the two methods. There appeared to be no significant differences in the quality and quantity of usability findings between remote and in-lab approaches.

TABLE 1.1   OVERVIEW COMPARISON OF LAB AND REMOTE METHODS

	Lab	Remote
Number of Users	8	8
Recruiting Method	Recruiting agency	Online live recruiting
Recruiting Duration	12 days	1 day
Testing Duration	2 days	1 day
Location	Pleasanton, CA	CA, OR, NY, UT
Avg. Session Duration	85.6 min	51.5 min
Total Key Findings	98	114
Approximate Cost	$26,000	$17,000
Deliverables	Report, highlight video	Report, highlight video, survey responses

Case Study: Lab vs. Remote (continued)

DETAILED COMPARISON OF METHODS

Tables 1.2, 1.3, and 1.4 break down the process for each of the recruiting, testing, and analysis phases, respectively. The left-hand column describes the lab study details; the right-hand column describes the remote study details.

TABLE 1.2 LAB VS. REMOTE RECRUITING

	Lab	Remote
Recruiting Channel	Third-party recruiting agency schedules users	Pop-up screener on the Web site
Selection Method	Agency selects users based on predetermined recruiting criteria	Researcher selects users based on responses to screener questions
Recruiting Pool	Local users are selected to avoid travel expenses	Visitors to the Web site
Duration	12 days	1 day

Recruiting for the lab-based study was outsourced to a professional recruiting agency (see Table 1.3). Ten users were recruited, screened, and scheduled by G Focus Groups in San Francisco, including two extra recruits in case of no-shows. Recruiting eight users through the recruiting agency took 12 days. Agency-assisted recruiting successfully provided seven test subjects for the lab study; the eighth recruit did not fulfill the testing criteria.

Recruiting for the remote study was conducted using an online pop-up from the software company's corporate Web site. The recruiting pop-up, hosted by the researchers, used the same questions as the G Focus Groups' recruiting screener. Users in both studies were selected based on detailed criteria such as job title and annual company revenues. Respondents to the online screener who met the study's qualifications were contacted in real time by the research moderators. The online recruiting method took one day and yielded eight users total from California, Utah, New York, and Oregon. Normally, the live screener requires four days of lead time to set up, but in this case it was completed for a previous project so setup was not necessary.

Case Study: Lab vs. Remote (continued)

TABLE 1.3 LAB VS. REMOTE ENVIRONMENT

	Lab	Remote
User Surroundings	User in controlled lab environment	User in native environment
Researcher Surroundings	On location with user	Testing globally distributed users from one location
User Screen Observation	Practitioner sees user screen on her computer	
User Observation	Through a one-way mirror	None
Communication Medium	Via microphone and speakers	Via telephone
Client Observation of Users	Through a one-way mirror; user audio, screen, and facial expressions are captured for later viewing	Live session observation via computer screen sharing; user audio and screen are captured for later viewing

The lab study was also conducted from the software company's in-house usability lab. The recruits for the lab study went to the lab in Pleasanton, California to participate and used a Windows PC. In addition to users' audio and screen movement capture, users' facial expressions were also recorded.

The remote usability study was conducted using a portable lab from the software company's headquarters in Pleasanton, California. The live recruits participated from their native environments and logged on to an online meeting allowing the moderators to view the participants' screen movements. The users' audio and screen movements were captured.

Case Study: Lab vs. Remote (continued)

The lab study uncovered similar issues of similar quality and usefulness to the client when compared with the remote study results (see Table 1.4). The remote study uncovered usability issues of high value to the client. The lab method uncovered 98 key findings, compared with 116 findings in the remote study (not a statistically significant difference).

TABLE 1.4   LAB VS. REMOTE FINDINGS

	Lab	Remote
Quality of Findings	High	High
Number of Usability Issues Uncovered	98	116
Video Deliverables	Highlights video with picture-in-picture	Highlights video with picture-in-picture, audio

The Case Against Remote Research

by Andy Budd

Even though we're obviously firm advocates of remote methods, not all UX practitioners agree. **Andy Budd** *is the creative director at Clearleft, a renowned London-based team of UX and Web design experts. Clearleft are the makers of Silverback, in-person usability testing software for interface designers and developers. Andy isn't such a big fan of remote methods for moderated studies—here's why.*

Full disclosure: I've done very little remote testing, and the reason is that I've never found a credible need to do it. We've always found other ways of testing that didn't require a remote approach. My issue is less about the negatives of remote testing and more about the positives of in-person testing.

Now, that's not to say we test in a *lab*; I find that labs give a veneer of formality and scientific accuracy to studies, which they often don't have. And testing labs and equipment are often more expensive, and tend to bog things down. So we take a grittier approach—just a meeting room and a video camera or some screen capture software.

The Case Against Remote Research (continued)

We gain a lot of information by being in the room with people. They say 90% of communication is nonverbal. It's about the cues in people's tone of voice or posture. When you're with a test subject, you pick up these signals more easily. With online video conferencing such as Skype, social conventions break down; you're not able to read the cues that tell you when one person stopped talking or when it's OK for another person to start talking. You get lag, and people talk over each other. Communicating remotely is difficult to do well, and it's possible that it's to do with our ability to use these new tools and technology. I wouldn't be surprised—give it 20 or 30 years—when video conferencing becomes a norm and we've learned how to understand and read these subtle cues better. For now, there's the potential to lose 90% of the information that's coming through to you if you're not testing in person.

ON THE SHORTCOMINGS OF REMOTE METHODS

Usability testing is all about empathy. It's about creating a connection. That kind of empathy is difficult to create through Web conferencing, and it's that gulf of miscommunication that makes it less attractive to me. I think there are instances where you should use remote moderated testing, often when it's impossible to recruit users to a specific location. Recently, we were working on a project for a South American site. We wanted to speak to Brazilians, so initially we thought to do some kind of remote testing, but then realized that there was actually a large student and ex-pat Brazilian community here [in Brighton]. So instead we went to a Brazilian café and sat down and just chatted to actual Brazilian people who happened to be living in the UK. Some people said, "How can you possibly say that talking to Brazilian people in a Brighton café is exactly the same as in a favela in São Paulo?" But again, the difference is so subtle as to make little difference on the results, particularly when testing in small numbers. Sure, if you're doing a scientific test and looking across very large sample sets, these minute effects are going to play a bigger role.

It also depends on what you are testing. There are some obvious cultural differences with the way people use the Web, but there are also universal habits: registering for a service, noticing positioning, etc. It's unlikely that the Brazilian community in Brighton is not going to pick up something the community in Sao Paulo might.

The Case Against Remote Research (continued)

ON THE USE OF TECHNOLOGY IN UX RESEARCH

People often try and find technological answers to human problems. A lot of the drive for remote testing is an attempt to find shortcuts. "Oh, it's difficult to find test subjects, so let's get technology to help us out." Today I was reading an article where someone was saying, "How can we do remote ethnographic research? Can we get live cameras streaming?" And I thought, "Do a diary study." Diary studies are probably the ultimate in remote ethnographic research. You don't need a webcam streaming back to Mission HQ. People love tinkering with technology because it makes them feel like superheroes; it's something to show off. I believe in human solutions; I think technology is often used as a crutch.

I think remote testing is still in its infancy. It's based on the technology that's available. It's preferable to have lightweight technology that you can send to a novice user, double-click on it, and it opens and installs. But if you're looking at an average computer desktop, which is now above 1024 × 768, and you also want to capture the person's reactions, you want to send audio and video down that pipe as well—that's a complicated problem. You need good bandwidth to do this really well. So then you create artificial problems, because you're limited to people who have got pretty good tech and bandwidth. And so that would probably prevent us going and doing remote testing with somebody in a cybercafé in Brazil.

What I *am* kind of interested in is unmoderated [i.e., automated] remote testing, because it's a hybrid between usability testing and statistical analysis or analytics. The benefit is that you can test on a much wider sample set. It complements in-person usability testing.

ON THE PURPOSE OF USER TESTING

The point is to develop a core empathetic understanding of what your users' needs and requirements are, to get inside the heads of your users. And I think the only way you can do that is through qualitative, observational usability testing. There are lots of quantitative tools out there, stuff that can tell you what's happening, but it can't necessarily tell you *why* it's happening. We've all done usability tests where you watch people struggle and have a real problem doing something, and you can see they're having a problem. Then they'll go, "Oh, yes. It was easy." We did a test where a user thought he'd purchased a ticket, and he hadn't, but he'd left thinking he had. If he'd told you, "Yes, I've succeeded," you would have been mistaken. Watching and observing what users do is very enlightening. Frankly, it's easier for people to learn from direct experience than through analyzing statistics. There's nothing like actually watching people and being in the same room.

Moderated vs. Automated

So, you've decided it's worth a shot to try a remote research study. Feels good, doesn't it? The first thing to know is that remote research can be roughly divided into two very different categories: moderated and automated research.

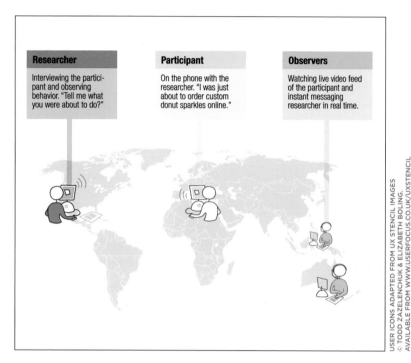

Researcher
Interviewing the participant and observing behavior. "Tell me what you were about to do?"

Participant
On the phone with the researcher. "I was just about to order custom donut sparkles online."

Observers
Watching live video feed of the participant and instant messaging researcher in real time.

USER ICONS ADAPTED FROM UX STENCIL IMAGES © TODD ZAZELENCHUK & ELIZABETH BOLING. AVAILABLE FROM WWW.USERFOCUS.CO.UK/UXSTENCIL

FIGURE 1.3
Moderated research: a researcher ("moderator") observes and speaks to a participant in another location. Outside observers can watch the session from yet a third location and communicate with the moderator as the session is ongoing.

In *moderated research*, a moderator (aka "facilitator") speaks directly to the research participants (see Figure 1.3). One-on-one interviews, ethnographies, and group discussions (including the infamous focus group) are all forms of moderated research. All the parties involved in the study—researchers, participants, and observers—are in attendance at the same time, which is why moderated research is also sometimes known as "synchronous" research. Moderated research allows you to gather in-depth qualitative feedback: behavior, tone-of-voice, task and time context, and so on. Moderators can probe at new subjects as they arise over the course of a session, which makes the scope of the research more flexible and enables the researcher to explore behaviors that were unforeseen during the planning phases of the study. Researchers should pay close attention to these "emerging topics," since they often identify issues that were overlooked during the planning of the study.

Automated (or "unmoderated") *research* is the flip side of moderated research: the researcher has no direct contact or communication with the participant and instead uses some kind of tool or service to gather feedback or record user behaviors automatically (see Figure 1.4). Typically, automated research is used to gather quantitative feedback from a large sample, often a hundred or more. There's all sorts of feedback you can get this way: users' subjective opinions and responses to your site, user clicking behavior, task completion rates, how users categorize elements on your site, and even your users' behavior on competitors' Web sites. In contrast to moderated research, automated research is usually done asynchronously: first, the researcher designs and initiates the study; then the participants perform the tasks; then, once all the participants have completed the tasks, the researcher gathers and analyzes the data.

FIGURE 1.4
Automated research: a Web tool or service automatically prompts participants to perform tasks. The outcome is recorded and analyzed later.

When to Use Which Remote Method

There's plenty of overlap between automated and moderated methods, but Table 1.5 shows how it generally breaks down.

TABLE 1.5 MODERATED VS. AUTOMATED RESEARCH

	Moderated	Automated
Timing	Synchronous—moderator speaks while participant uses Web site	Asynchronous—participant performs task, researchers analyze the results later
Sample Size	Small—about 5 to 30, depending on user segmentation and study goals	Large—usually 50 or more
Incentives	Usually ~$75 per participant, but can vary depending on who's participating	Can offer 1 in 10 participants a normal incentive (~$75) or a smaller incentive (~$5) to all participants
Types of Research Goals	Discovering errors and usability issues, formative research, Time-Aware Research, real-world contextual research	Assessing performance on narrowly defined tasks
Type of Research	Qualitative and behavioral, but can also ask opinions and gather rich contextual info	Quantitative, both behavioral and opinion-based, but can also ask brief open-ended questions—less context

Moderated research is qualitative; it allows you to observe how people use interfaces directly. You'll want a moderated approach when testing an interface with many functions (Photoshop, most homepages) or a process with no rigid flow of tasks (browsing on Amazon, searching on Google) over a small pool of users. Since it provides lots of context and insight into exactly what users are doing and why, moderated methods are good for "formative research" when you're looking for new ideas to come from behavioral observation. Moderated research can also be used to find usability flaws in an interface. We cover the nitty-gritty of remote moderated research in Chapter 5.

Automated research is nearly always quantitative and is good at addressing more specific questions ("What percentage of users can successfully log in?" "How long does it take for users to find the product they're looking for?"), or measuring how users perform on a few simple tasks over a large sample. If all you need is raw performance data, and not *why* users behave the way they do, then automated testing is for you. (Suppose you just want to determine what color your text links should be: testing every different shade on a large sample size to see which performs best makes more sense than closely watching eight users use three different shades.) Also, some automated tools can be used to gather opinion-based market research data as well, so if you're looking for both opinion-based and behavioral data, you can often gather both in a single study. And certain conceptual UX tasks, like card sorting and A/B testing, are well supported by automated tools. See Chapter 6 for a more thorough look at the various automated research methods.

You don't necessarily have to choose between moderated and automated testing, or even between lab and remote methods. You can even conduct multiple studies on the same interface, using the findings from one study to add nuance to another. That's probably excessive for the average study, but for really large-scale projects where you just want to gather every bit of information you can (a new version of a complex software program, an overhauled IA, etc.), being comprehensive can't hurt.

After reading this chapter, you should have a good idea of whether or not remote research suits you. Give it a try—if it's not your thing, you can always go back to lab testing. We won't tell anyone.

Chapter Summary

- Do a lab study when you need to use special equipment, keep the interface 100% secure, see the user's physical movements, or when you can't use screen sharing tools.

- Remote research has its own strengths, the greatest of which is that it enables Time-Aware Research, in which you observe users performing tasks you're interested in observing right at the moment they'd naturally perform them.

- Remote methods can also give you greater user diversity, cut travel costs, allow you to test from anywhere, be quicker to set up, and can be used to test context-dependent interfaces that wouldn't make sense in a lab.

- Currently, remote methods work best for testing functional computer interfaces, prototypes, wireframes, design comps, and mockups. You can test physical products with the right setup, but it's slightly harder.

- There are two kinds of remote research: moderated and automated. Moderated research has the researcher communicating with and observing users as they perform tasks; automated research has researchers using online tools and services to collect behavioral data from users automatically.

- Generally, moderated research is good for collecting rich, qualitative behavioral data from a small sample, while automated research is good for collecting quantitative data over a larger sample.

CHAPTER 2

Moderated
Research: Setup

L et's get down to setting up a typical moderated one-on-one study. We'll walk you through gathering all the equipment and software you'll need and explain how to prepare for your first research session. To keep things moving, we'll stick to bare-bones basics in this chapter, explaining the simplest way to set up a generic moderated research session. Later, you'll learn about other tools, approaches, and strategies you can use to develop a study that best suits your particular needs.

Gearing Up: Physical Equipment

Even though you don't need a lab to do remote research, you'll still need some equipment to make calls, see your users' screens, and record the sessions, and there are also a few tools that can make your life easier. Fortunately, you can find a lot of these lying around most offices (see Figure 2.1).

FIGURE 2.1
A standard remote lab setup: a two-line desk phone, laptop with wired Internet connection, a second monitor, and a phone headset. You may also need a phone tap and amplifier to record audio, depending on your recording setup.

- **A computer.** Laptop or desktop—doesn't matter. Just be sure that the system specs exceed the minimum requirements of all the screen sharing, chat, recording, and note-taking software you'll be running. And be especially sure that your computer is compatible with the screen sharing solution you choose; lots of screen sharing solutions are currently PC-only, although this will probably change. And be doubly sure to have plenty of hard drive space free.

- **A wired, high-speed Internet connection.** Wireless connections are too unreliable to run screen sharing software with 100% confidence. It's very important to have an uninterrupted wired connection—the faster, the better.

- **A landline, two-line, touchtone desk phone.** It has to be a landline desk phone, for a few reasons: batteries won't last across multiple testing sessions, a phone tap (if you need one for recording) won't work with a mobile or cordless phone, and most importantly, it's more stable and reliable. And it has to be a two-line phone if your setup requires you to conference call with both your user and your observers. (Alternatively, you could use a VoIP service like Skype if you're prepared to deal with the spottiness and instability of the typical Skype call, circa 2009.)

- **Headset for the phone.** Your neck will thank you.

- **Two monitors.** The screen sharing window alone takes up nearly an entire monitor, so if you actually want to be able to see your notes, chat windows, or anything else on your computer, you'll need the screen space.

- **Phone recording adaptor or a phone tap and amplifier.** You'll need these tools to record the phone conversation if you're speaking through your landline telephone (and not a VoIP service).

- **Peace and quiet.** It's crucial to test in a place where you can talk freely at a normal volume and won't be disrupted, like an empty office or meeting room. Background noise can be unbelievably distracting to both the moderator and participant.

Doing a Pilot Test Right Now

So now that you've got your equipment together, why wait? Let's do a simple 10-minute pilot test to get your feet wet. This pilot test will simulate a basic moderated session, not including the recruiting process (which is described in the next chapter).

The crucial piece of software you'll need is a screen sharing application, which will allow you to see what's on your user's computer screen during the session. There are many screen sharing options (described at length in Chapter 8, "Remote Research Tools"), but for now we'll stick with Adobe's Acrobat Connect. It supports observation and chat, as well as webcam sharing (which we won't get into here). Connect doesn't require users to install anything on their computers; all they have to do is visit a Web site that you'll direct them to, which means that you can get around most firewalls, antivirus software, and other barriers that might prevent you from running the screen sharing. It's compatible with all OS platforms, so you, your observers, and your participant can be on a Mac, Linux, or PC. And finally, it comes with an optional conference-calling service that you can use to have the study participants and observers on the same line.

Now, you'll need a pilot participant. Grab anyone at all—your coworker, sister-in-law, high school lacrosse coach—and just tell him/her in advance to be waiting near a phone and a computer with high-speed Internet access.

OK, time to get connected with your participant. For brevity's sake, we'll assume you can follow onscreen instructions:

1. Sign up for a free trial of Adobe Connect at the Adobe Web site (google "Adobe Connect") and follow the instructions there to begin a session. You should end up at the Connect session window (see Figure 2.2). The first time you use Acrobat Connect, click the Share My Screen button. That will trigger the download of a little plug-in that will allow you to use the screen sharing function. (This is the only time you'll have to do it; after that, it should always launch using the plug-in, rather than a tab of your Web browser.)

2. Call your pilot user.

FIGURE 2.2
The Adobe Acrobat Connect session window: the big window is the user's screen, and the smaller windows, from top to bottom, are the user's webcam (optional), the participant list (including observers), chat, and notes.

3. Set up your user's screen sharing. Tell your user to start the screen sharing session by going to connect.acrobat.com/XXXXX (again, XXXXX = your Connect account name) and join as a guest. (If you're tech-savvy, you can shorten the URL using http://tinyurl.com or set up an easy-to-type redirect link in advance and tell your user to go there. When you're reading a URL over the phone, it's easy for listeners to get the letters confused, so the shorter, the better. It's usually faster to do a careful "T-as-in-Tom, O-as-in-Orange" spelling to avoid errors.) Tell the user to click on the arrow in the upper-right corner of the Connect window to download the screen sharing plug-in and then tell him/her to click on the Share My Screen button that appears. You should be able to see the user's screen in your session window now.

4. Begin the study! It's just a pilot test, so do whatever you like: have the user show you how he/she uses your interface, watch him/her fill out a survey...anything. When you're done, click the End Session button in the session window, and the screen sharing will end. No uninstalling

is necessary for Adobe Connect. Here, you should practice the things you say to the participant when wrapping up the study, the most important being: "Thank you *so* much for participating, I really appreciate it," and "I can no longer see your screen and will not be able to do so again." We'll cover more of these kinds of details later in the chapter. For now, you're all done.

Preparing for a Real Study

So now you know what you need for a basic study: computer, high-speed Internet, phone, and some screen sharing software. Now let's back up and talk more in detail about the function of each tool.

Screen Sharing

Usually, the hardest part of remote research is getting your users to successfully share their screen with you—that includes both obtaining their consent and getting them to set up the screen sharing tool. As with so many technical pursuits, the more attention you give to the setup ahead of time, the easier your life will be when you actually start calling people.

First, know this: about 15–30% of all the remote sessions you attempt to set up will fail for one reason or another (see Chapter 10, "The Challenges of Remote Testing," for more details on the kinds of challenges you'll encounter). That's just the wild world of the Internet, and this is why a lot of the articles on remote testing focus on the nuts and bolts of conducting sessions; a few years will have to pass before the technical details become less of a pain.

Choosing the right screen sharing tool can be overwhelming when you're not familiar with the pros and cons of each tool. Most practitioners who do lots of remote testing eventually choose one tool and stick with it, but you should try a couple before figuring out which one works for you (most offer free trials). Broadly, the most important considerations for a screen sharing tool are its compatibility and ease of setup for both the moderator and the participant. Browser and OS compatibility have a big impact on whom you'll be able to talk to, and you probably don't want to arbitrarily limit your recruiting pool by who can run your screen sharing software (unless the interface you're testing happens to be platform specific, too, in which case it doesn't matter). You want to make the setup process as quick and smooth as possible so that you don't prolong the session or frustrate the user with tedious instructions.

For platform versatility, you can't beat browser-based screen sharing solutions like Connect, GoToMeeting, and LiveLook, since they only require users to run a fairly recent browser, and the OS doesn't matter. The only minor downside is that these solutions currently require users to have a Java-enabled browser; most are, but if some users don't have it, having them set it up can be time-consuming, and they may not even want to.

The solutions that aren't browser based are often the ones that require users to download and run an executable file or have a certain program installed (Skype, iChat). Downloading and installing files can be a security issue for users who are behind corporate firewalls, as well as a trust issue for users who don't like the idea of installing things at the command of a voice on the other end of a telephone. If you decide to use screen sharing software that requires any heavy installation or user registration, you'll probably want to arrange in advance for your users to do that.

If you plan on having people observe your sessions remotely, screen sharing features may make that a lot easier. Most tools limit the number of observers; make sure that yours can support as many as you need. A handful of tools enable chatting between the observers and moderator, which is handy, but the participant should always be excluded from the chat, to keep distraction to a minimum. If it's not possible to block the participant from the chat, you can use an instant messaging service to chat with clients behind the scenes.

On a final practical note, there's cost to consider. Pricing structures differ from tool to tool but generally fall into a handful of categories. Adobe Connect and GoToMeeting offer subscription plans, which are best if you intend to conduct multiple usability testing sessions over the development of your interface (which we encourage), and they also offer free trials if you just plan on doing a one-off study. Tools like Skype (for international calls) charge just a few cents a minute, which is cost effective if you don't plan on conducting usability studies regularly, but they make you buy a set number of minutes in advance. Watch out for additional charges you'll incur if the screen sharing lacks a particular feature you need; for instance—LiveLook currently lacks support for observers to listen in on sessions, so you'd have to pony up for a conference call service, which can be pricey. And if you're using your phone to dial internationally instead of a voice chat service like Skype, watch those long-distance charges.

For the full story on the state of remote research tools, check out Chapter 8. For now, Table 2.1 provides a quick comparison of some handy screen sharing solutions.

TABLE 2.1   SCREEN SHARING TOOLS AT A GLANCE (CIRCA MID-2009)*

	Adobe Connect	GoToMeeting	LiveLook Assistant	Skype 2.8
Can I see the user's whole desktop?	Yes	Yes; can also restrict to single window	Yes	Yes
Can it record the session?	Yes (with Pro version)	Yes	No	No
Does it support observers?	Yes	Yes	Yes (no audio)	Yes
Does it support chat?	Yes	Yes (visible to participant)	No	Yes
Platform	PC, Mac, Linux	PC, Mac	PC, Mac, Linux	PC, Mac, Linux
Pricing	$39.95/mo for standard, $375.00+/mo for Pro	$49.00/mo, $468.00/yr, 30-day free trial	$360.00/year	Free domestic; variable international rates
Does the user have to install anything?	No; plug-in download required	No; Java-enabled browser required	No; Java-enabled browser required	Yes; must have Skype installed and have access to Skype account
Can it share multiple monitors?	No, but you can choose which to view		No, primary monitor only	

* See Chapter 8 for more details.

Recording

Session recordings are useful if you want to be able to document, share, and closely analyze the testing sessions after the fact (bear in mind, though, that recording is not strictly essential for the purposes of running the study). Unlike in-person testing, in remote testing you can usually capture everything important about a session with just a simple software tool. Some screen sharing tools, like GoToMeeting and UserVue, have built-in recording functionality, and some Skype plug-ins will record Skype video as well (for example, eCamm's Call Recorder for Mac). If recording functionality isn't built in, you'll need a screen recording application that allows you to capture both video and audio from your computer. Techsmith's Camtasia Studio is a common one, although there are free applications like CamStudio as well. Most recording applications produce either WMV or AVI files. Recording is probably the most computing-intensive operation on the moderator's end, especially if you're recording a large screen area, so be sure that whatever you use to record doesn't noticeably slow your computer.

Taking Notes

While some people are still most comfortable with handwritten notes, there are advantages to taking notes on the computer during remote testing. For one thing, you don't have to take your eyes off the screen to take notes, and for another, most people (most UX professionals, anyway) type faster than they write. As with any user research, you'll be furiously taking notes on what users say and do, as well as things you see on their screens incidentally. Fifty words per minute is close to the speed you'll have to type if you want to transcribe user quotes verbatim, but don't get distracted from what's happening in the session; you can resort to your session recordings later, if you have to.

A good old word processing document or Excel spreadsheet should suffice for basic note taking, but if you want to make analysis easier, you can find ways to cleverly take notes that are automatically timestamped to the video recording you're making. Timestamping your notes (i.e., keeping track of the exact time you took each note) matters only if you're planning on making highlight videos or conducting detailed analysis after your sessions are complete. For now, we stick to basics; see Chapter 5, "Moderating," for more about note-taking and transcription techniques and Chapter 7, "Analysis and Reporting," for more on the analysis process.

Webcams

We don't ordinarily use a webcam in remote usability sessions, but it's an interesting option. It allows you to see your users' facial reactions during the session or vice versa. A growing number of screen sharing applications have webcam functionality built into them. This, of course, would require your participants to be using a webcam, and they'd also need to have any necessary software. Adobe Connect has an integrated screen sharing/video conferencing solution, Skype has both video chat and screen sharing, and Google just recently introduced webcam chat built right into its Gmail chat client.

As with recording, you need to make sure that any webcam software isn't going to kill your computer performance; recording is CPU intensive, so test it in advance. And, as with any personally identifying information gathered during a remote research study, make sure that images and recordings of users are gathered with proper consent (see Chapter 4, "Privacy and Consent") and securely stored or erased, to respect their privacy and abide by all relevant laws.

Drafting the Research Documents

As we mentioned in the introduction, this book focuses on how to do *remote* user research, so we assume you know how to plan and manage a standard user research study. Specifically, we're assuming you know how to schedule the project, define the research goals and user segments, recruit users (if you're recruiting the traditional way), and get everyone (researchers, stakeholders, observers) on the same page about what needs to be done. (If you have no idea what we're talking about, see Chapter 5 of Mike Kuniavsky's *Observing the User Experience*.)

There are a few ways in which preparing for a remote study is different. First, the facilitator guide (aka "moderator script") will contain important new parts that cover introduction, screen sharing, user consent, and incentive payment. Second, the observers need to be briefed on how to use the screen sharing service and how to communicate with the moderator. Finally, if you're recruiting on the Web, you'll need to design a recruiting screener and place it on your Web site in advance. We'll get into that topic in the next chapter.

The Remote Facilitator Guide

Drafting the facilitator guide is usually the most time-consuming part of the research setup process. Note that we deliberately call it a "guide" and not a "script"; you will not be mindlessly reading verbatim from this document. The facilitator guide should always be written with flexibility in mind, anticipating the unpredictable things that happen when you do time-aware research. You're speaking with people who are in their own native environments performing natural tasks, outside a controlled lab setting. Think of the facilitator guide as a document that establishes everything you should expect to encounter over the course of the session but shouldn't restrict you from exploring new issues that come up.

A facilitator guide divides your study into four main sections, totaling about 40 minutes: Greeting and Introduction, Natural Tasks, Predetermined Tasks, and Wrap-up and Debrief. Since studies tend to run long, we've found that 60 minutes is about the upper limit for maintaining a participant's attention and investment in the study.

Presession Setup (1 Minute)

Set up a screen sharing session and run your recording software. This method will vary depending on what screen sharing and recording tools you're using.

Cue your observers to join the session. If you have people observing the session, cue them to join the session before contacting the user. (See "Preparing Observers for Testing" later in this chapter for more details.)

Greeting and Introduction (5 Minutes)

The first part of the guide will deal with establishing contact with the participant. This part comes closest to resembling a "script" in the traditional sense; most of what you need to say probably won't change much from user to user. The introduction in our sample script has been refined over hundreds of sessions and is the most efficient way we've found to introduce a study.

We won't be reprinting a full facilitator guide here (although you'll be able to find examples on http://remoteusability.com), we want to point out a few of the important things all facilitator guides need to establish. Elements that will vary from study to study are highlighted in yellow, both here and throughout this book:

Contact and self-introduction. Contact users. Right away, introduce yourself with your name, the company or organization you represent, and remind users how you got their name and phone number. If you're live recruiting (i.e., calling users who just filled out a Web recruiting form, see Chapter 3, "Recruiting for Remote Studies"), you have to do this pretty quickly. Otherwise, they might mistake you for a telemarketer.

> Hello, can I speak to Bill Brown? Hi Bill. My name's Nate, and I'm calling on behalf of ACME about the [usability study we had scheduled for this time] / [survey you filled out at ACME.com a few minutes ago].

Willingness to participate. Confirm that users are available and willing to speak for the duration of the study.

> Do you still have time now to talk to me for 40 minutes about the ACME Web site? *[If not, ask whether you can reschedule, and then end the call.]*

Ability to participate. Confirm that users have the necessary equipment to participate in the study. You'd be surprised how many people aren't able to talk on the phone and use their browser at the same time.

> Will you be able to use Firefox and talk on your phone at the same time? And do you have a high-speed Internet connection? *[If not, end the call.]* Great!

Obtain consent for screen sharing and recording. This part takes some finesse, because if your users didn't anticipate the call (which is always the case with live recruiting), they may understandably become suspicious when you, a stranger on the telephone, ask them to download something. Explain in clear, simple language what you'd like them to do, why, and what they're getting themselves into. If you're using a Consent Agreement (as described in Chapter 4), direct the users to the consent form here.

> So, during this call today, we'd like to follow along with what's on your screen while we're talking to you, and to do that, we ask you to visit a Web site that will allow us to see and record whatever you can see on your desktop. The recording is used only for research, and the screen sharing is completely disabled at the end of the session. Does that sound okay? *[If not, end the call.]* Great!

> *[If using a Consent Agreement]:* I'd like to direct you to a Web site with a consent form that describes what the study will be about, so you can make your participation official. The address to put in your browser's address bar is: www.acme.com/consent.

Screen sharing setup for the participant. The details of this step depend on the tool used.

Introducing the study. Establish the users' expectations about what will happen during the study and what kind of mindset they should have entering the study. The most important things to establish are that you want the participants to use the interface like they normally would:

> So let me tell you a little bit about what we'll be doing today. Basically, I'm going to be getting your feedback on ACME.com. Your job today is going to be really easy. Basically, I want you to just be yourself and act as you naturally would on the Web site. Now and then, I may chime in to ask you what you're thinking, but mostly I'd just like you to show me how you use the site. If there's a point where you'd normally quit or stop using the Web site, just let us know.

And let them know you'd also like them to think aloud while they're on the site:

> There's just one thing that I would like you to do differently, which is to "think aloud" as you use the ACME Web site. For example, if you're reading something, read it aloud, and feel free to comment as you read—for example "that's interesting" or "what in the world are they talking about?" Let me know what's going through your mind while you do things, so I can understand exactly what you are thinking and doing (for example, "Now I'm going to try to use the search engine"). If you get to a point where you would naturally leave or stop using the Web site, let me know.

If your interface is a design in progress, you should set those expectations early so that users don't bother wasting time trying to figure out something that isn't finished anyway. You should remind them, however, that there's no need to modify their behavior just because it's a prototype. Our experience is that most people can't tell the difference between a black-and-white prototype and a real functioning application anyway.

> Also, keep in mind that some things might not be working today on this prototype, which is fine. If you run into something that doesn't work, I'll let you know, and you can tell me what you were trying to do.

It's also nice to set users at ease by reassuring them that you had nothing to do with the design of the interface, so they can be completely honest:

> And finally, I want to let you know that I had nothing to do with the design of the Web site. I'm just collecting feedback, so feel free to be candid with your thoughts. No need to worry about hurting my feelings or getting anyone in trouble. Does that sound good? Great! Then let's move on to the Web site.

Natural Tasks (15–25 Minutes)

If you're conducting time-aware research, for the most part you won't have to explicitly assign tasks. For example, if you're testing the login process for your Web site, you can contact people just at the moment they first arrive at your homepage, when you know they haven't yet logged in. This is what we call a "natural task"—something users were going to do anyway, whether or not they were participating in a study.

Since natural tasks have a lot of variability, this section of the script should be extremely open-ended. The moderator's comments should be geared toward encouraging the user to do what he was intending to do with the interface, *without* directing the user to perform any specific tasks[1]. This is often the most important part of the testing session, but it's actually the shortest part of the script simply because you can't predict in advance exactly what the users will do. You may include some typical prompt questions, for your own reference, such as the following:

- So, tell me what you're looking at.

- What's going through your mind right now?

- What do you want to do from here?

- Where would you go from here?

- Which parts of this page seem most interesting to you?

- What kind of info are you looking for on this page?

- What stands out to you most?

- What were you expecting when you first came to this page?

- When did you decide to leave the site/exit the program?

- What do you think about the way everything's laid out?

- What brought you to this page?

- Is there any more info on this page you'd like to see included that you don't already see here?

---

[1] Chapter 5 covers the finer points of moderating in detail; for more details on traditional moderating techniques, see Joseph Dumas and Beth Loring's *Moderating Usability Tests: Principles and Practices for Interacting.*

Predetermined Tasks (5–15 Minutes)

Since natural tasks are unpredictable, they may not always cover every last issue you're curious about, so you'll want to make sure you're prepared to address any tasks, questions, or parts of the interface required by the study goals. This will require you to write down *all* the important questions you may have, even though you won't necessarily be asking them all. Ideally, you want to cover as many of those questions as possible in the natural tasks section. Even though this section will appear long in the guide, you won't be asking every question; therefore, it shouldn't actually take as much time as the natural tasks.

Wrap-up and Debrief (3–5 Minutes)

Concluding the session should be the quickest part of most studies. Participants should be explicitly directed to exit or uninstall any screen sharing tools and tracking software used during the session and also informed about how they will receive incentives for participating in the study. It's good to give participants a way to contact the researcher, in case they come up with any other questions regarding the study. And don't forget to say thank you!

Well, Bill, that does it. Let me just verify the email address where you would like us to send your incentive. *[Verify necessary contact info.]* The Amazon Gift Certificate usually takes 1–2 weeks to arrive and will be delivered via email. Check your junk mail folder if it doesn't show up, and if you don't receive it 2 weeks from today, send me an email at moderator_email@acme.com.

> *[Disable screen sharing.]* I've removed the screen sharing plug-in, so no need to get rid of that yourself; it's completely removed from your computer.
>
> Do you have any last questions for me? Thanks so much for talking to me today; it's been really helpful for us. Have a good day!

Preparing Observers for Testing

If you have colleagues or clients who plan on observing the sessions, you'll have to give them their marching orders. Observers need to be aware of times when the sessions will be happening, how to use the screen sharing tool, and how to speak directly with the moderator during the session (which is one of the perks of remote testing).

A standard way to prepare observers is to send an email with a complete set of instructions one week before testing and then a follow-up reminder email the day before testing, with the same set of instructions appended. Our sample instruction email shown in the sidebar assumes you're using Adobe Connect to do screen sharing and using live recruiting techniques, but we have instructions for other tools posted at our Remote Usability Web site (http://remoteusability.com).

On the day of testing, observers should be at their computers during testing hours, probably with a set of headphones so that they don't have to hold the telephone all day long. You should let the observers know that depending on the scheduling and the method of recruiting, there may be significant downtime between each session. If you're using live recruiting methods (as described in the following chapter), the amount of time between sessions may not be predictable, depending on the number of recruits you're able to obtain.

One thing to note: if you're using live recruiting techniques, you won't know exactly when the testing sessions are going to be held, so your email will have to set observers' expectations that there may be periods of downtime between sessions, so they should have other things at hand to keep busy with.

Observer Instructions

The following is a sample email you should send to observers at least a week in advance of testing. The main point of the email is to inform observers what they need to observe in the sessions, how to set up the screen sharing tools, and what the ground rules are for communicating with the moderator. You want to encourage observers to contribute helpful observations and advice to the moderator during the session, but advise them not to flood the moderator with distracting messages and have them generally defer to the moderator's judgment in matters of running the session.

This letter assumes you're using Adobe Connect, but other variations can be found on our Web site at http://remoteusability.com.

Hi everyone,

This email contains detailed instructions for how to observe the remote user research sessions, which will begin next week.

[How to Set Up Adobe Connect]

For most of the screen sharing with our participants, we'll be using a Web service called Adobe Connect. You don't have to install anything now, but you will have to follow certain steps before each individual session. Here's how you'll be using it:

1. **Go to the Web address** at the beginning of the test session: connect.acrobat. com/XXXXX

2. **Enter as a guest.** Use a neutral guest name like "usability" or "user testing" so that the participant isn't distracted by seeing an unfamiliar name.

3. **Dial in to the conference number:** (XXX) XXX-XXXX

4. **Mute your phone!** This is very important; otherwise, the user will be able to hear you, which will disrupt the session.

Observer Instructions (continued)

[OTHER THINGS TO NOTE]

—Recruiting and testing will be conducted between the hours of Noon EST (9 AM PST) and 8 PM EST (5 PM PST), with a brief break around 3 PM EST so the researchers can get some lunch. Please be aware that for setup and recruiting processes, there may be several minutes of downtime between each session, so feel free to have something to work on while you wait for a new session to begin.

—Phone headsets are encouraged so that you don't have to hold the phone with your shoulder all day.

—I can be reached on AOL Instant Messenger at "aimscreenname." To keep from swamping me with requests from multiple people at once, please designate one person who will direct any questions and comments to me as the sessions are ongoing. Feel free to point out any and all issues you think are important. I may not be able to handle all your requests, but I'll try my best!

—If you have any other questions about how to follow along with the sessions or communicate with me during the sessions, feel free to email me.

That's all! Looking forward to getting this study started!

NOTE RESEARCHERS AND STAKEHOLDERS

In this book we occasionally mention "researchers" and "stakeholders" because oftentimes the people conducting the research ("researchers") are doing it on behalf of other people who have commissioned the research ("stakeholders"). Stakeholders— who can include business executives, managers who dictate the research budget, and so on—are usually untrained in research methods (never mind *remote* research methods), but they're essential for defining the parameters of the study and should be involved in all steps of the testing. Of course, there are also situations in which the researchers *are* the stakeholders—e.g., academic research or companies in which the developers also do their own user testing. So, just to be clear: there aren't always business stakeholders, and in those cases you can ignore the whole researcher/stakeholder dichotomy and the issues of "sign-off and approval" we occasionally bring up.

Chapter Summary

There's always the chance you may have to go back and adjust something in the project goals or adjust the schedule for whatever reason; however, you can consider yourself pretty much finished with the setup phase and ready to move on to recruiting once you've completed the following:

- Selected, installed, and familiarized yourself with a screen sharing tool (and recording tool, if necessary).

- Set up your equipment: computer, high-speed wired Internet connection, monitor(s), two-line desk phone, phone headset, and any other necessary equipment.

- Arranged backup screen sharing tools, as well as any optional tools (recording, note taking, etc.).

- Completed standard user research project management steps (check out Chapter 5 of Mike Kuniavsky's *Observing the User Experience*).

- Completed the recruiting screener design and recruiting test (covered in the next chapter).

- Drafted and familiarized yourself with the facilitator guide.

- Briefed observers on how to participate during testing days.

- Conducted a practice run of the test, to make sure everything's working (especially if it's your first time).

Got all that? Then you're ready to start testing!

CHAPTER 3

Recruiting for Remote Studies

Recruiting research participants is notoriously frustrating and easy to mess up. If you *do* mess it up, you risk blowing the validity of the study, and even if you *don't* mess it up, it can still be a big drain on time and money. Many researchers are glad to hand over recruiting to a professional third-party recruiting agency, paying anywhere from $100 to $800 per recruit (depending on the stringency of the recruiting criteria). Others rely on in-house email lists, academic volunteer pools, paid participant panels, or personal contacts, each of which may not provide a representative or unbiased sample, for a number of reasons. And worst of all are online classified ads like those on craigslist, where the recruits are usually biased or, at best, solely interested (as opposed to *mostly* interested) in collecting an incentive check.

But then there's the rest of the Web: a huge pool of anonymous, disinterested, ordinary people who wouldn't necessarily consider participating in a research study, much less join a panel of standby participants—in other words, very promising research participants. This section will teach you how to recruit those people, confirm that they're qualified, contact them, and convince them to take 40 minutes out of their day to participate in a research study, all in a reliable, ethical, and nonirritating way.

Note that you can use these methods to recruit for *any* kind of study, whether moderated or automated, in-person or remote, live or scheduled. And we'll also explain why doing it this way is worthwhile.

What's "Live Recruiting"?

Live recruiting is using your Web site to collect voluntarily submitted user info and then using that info to contact qualified users within seconds. Why would you want to do that? First, it eliminates the need to schedule users in advance. Since you're using remote methods, you can begin a study at the very moment the recruit agrees to participate. And by intercepting visitors to your Web site using a form or pop-up window, you can instantly screen and call them within minutes of their submitting the form—that's what makes it "live."

Predictably, live recruiting has its advantages and limitations. The biggest advantage, which is an advantage of remote research in general, is that it enables time-aware research.

> **NOTE** **LIVE RECRUITING IS THE KEY TO TIME-AWARE RESEARCH**
>
> Remember that "time-aware research" concept we keep bringing up? Live recruiting is the easiest way to make it happen. By recruiting participants just as they're about to perform a task you're interested in watching, you can contact them and begin a remote research session right away. We can't overemphasize how much of a difference this makes.
>
> Let's say you want to study people who are browsing for laptops on your Web site. If you're live recruiting, you can wait until you hear from people who say, "I'm browsing for laptops," and then call them right away, so you can actually watch them browse for laptops on their own initiative. You don't have to tell them to pretend that they're browsing for a laptop. Users with real motivations for their tasks provide you with insights into how people actually use your product or Web site, which you can use to make your designs easier, better, more useful, and inspired by real behavior.

Another benefit to live recruiting is that it can ease the problem of participant flakiness. Tardiness and no-shows are a constant pain in prescheduled research studies because users can be late for a million reasons: couldn't find the lab, got stuck in traffic, just plain forgot, and so on. For every four to eight users who are scheduled, most recruiters arrange for at least one backup user, who must still be paid the full incentive amount. This problem disappears when you recruit live because the sessions aren't prescheduled.

Instead, with live recruiting what you need to worry about is having a constant stream of users responding to your recruiting screener. Many factors feed into this (discussed in the following section), but if your circumstances favor a high response rate, live recruiting has the potential to make recruiting cheaper and more reliable (see Table 3.1).

TABLE 3.1 COMPARING RECRUITING METHODS

	Live	Agency	Email/Personal Contacts
Cost	Cost of developing the screener or using a Web service (usually cheap)	Typically, $150–$600, depending on allotted time and criteria strictness	Free, but can be time consuming
Validity	Depends on screening ability of researcher	Depends on screening ability of recruiting agency, in cooperation with researcher	Low
Reliability	Depends on Web traffic; high, with 1,000+ unique visits per hour	High, but requires backups in case of no-shows	Low, depends on size of contact list; gets harder with subsequent studies
Time	Fast with 1,000+ unique visits per hour; not viable with fewer than 500 daily visits	1–4 weeks	Depends on size of available contacts
Effort	Medium; researcher bears responsibility for proper screening, participant data management	Low; researcher drafts requirements document	Medium; researcher bears responsibility for proper screening
Requirements	Healthy Web traffic, administrative/editorial access to a Web site	Enough money, a recruiting agency that covers local participants	An existing list of contacts you have permission to contact

Live recruiting also incorporates much of the recruiting phase into the testing phase, shortening the overall project schedule. Old-school recruiting methods require two to four weeks of contacting prospective users, screening and vetting them for their qualifications, briefing them on the study preliminaries, and sending reminders. And then once people start showing up, you're stuck with what you get. When you recruit live, the bulk of the screening is done through the Web form or within the first few minutes of calling a participant. This usually means you'll have to contact a few respondents before you can find one that's qualified, willing, and available to participate. The average time per session goes up, but if you have enough respondents, this approach is faster than advance screening.

NOTE CAN I RECRUIT FIRST-TIMERS?

Many of our clients want to know if it's possible to live recruit users who've never seen or had any experience with the product or Web site; the answer is *yes*. To do so, you need to have a screener question that determines whether or not the respondent has had any experience, and then you need to contact the respondent to begin the session *immediately* after receiving his or her response. Usually, you can catch the visitor within a minute of entering the site for the first time, which is fine for a "first-time visitor" for most purposes.

If you need a 100% new visitor, however, the best approach is to place the recruiting screener on a different Web site and then direct the recruit to the interface you want him/her to test. You may find, however, that you get a lower proportion of qualified recruits this way.

Finally, when you recruit from the Web, you're the one in charge, rather than a third-party agency, and your participants (except fakers, who are discussed later) will come from a single source—visitors to your Web site. This gives you more control and greater transparency over your participants, which can help boost the validity of the study. The recruiting sources of recruiting agencies aren't always clear, and often include personal contacts, standby participant panels (i.e., professional participants, often used in market research), and in some cases, even shady email lists. That doesn't necessarily translate into bad participants, provided the agency does its job right. With live recruiting, there's more transparency to the source of your recruits, and they're usually people who are coming to your site because they're genuinely interested in your products, services, software, cute puppy videos, or whatever.

NOTE REMOTE TESTING VS. REMOTE RECRUITING

Live recruiting is great for gathering participants for remote research studies; however, you don't *have* to use live recruiting for a remote research study, nor do you have to use traditional recruiting methods (agencies, email lists, panelists, friends, and family) for lab research. **Remote testing and remote recruiting are separate.** You can mix and match approaches to your testing and recruiting: use remote research tests with pre-scheduled participants or recruit people from your Web site for in-person lab tests.

Live Recruiting Using Forms and Pop-ups

There's lots of room to innovate with live recruiting. The one strict requirement is that you have administrative/editorial access to a Web site with a decent amount of traffic. (If not enough people visit your site, or you're still in super-secret startup mode and don't have a live Web site, you may be hosed and are probably better off going with a traditional recruiting method.)

Implementing the Recruiting Screener

The most straightforward option is to create a separate standalone page containing the recruiting screener form, which users can fill out to opt-in to your study. It's easiest to make by using an HTML form-building tool like the excellent Wufoo or Google Docs' form functionality (see Figures 3.1 and 3.2). You can then link to the standalone page from one of your main pages. Alternatively, if you're able to just embed the form somewhere on an existing page, users might be more likely to fill it out. Some remote research tools and Web services[1] like UserZoom and WebEffective offer JavaScript-based intercept forms as part of their comprehensive user research services but require you to sign up for the whole enchilada.

---

[1]   We maintain a list of services on our Remote Usability Web site (http://remoteusability.com), as well as in Chapter 8 of this book.

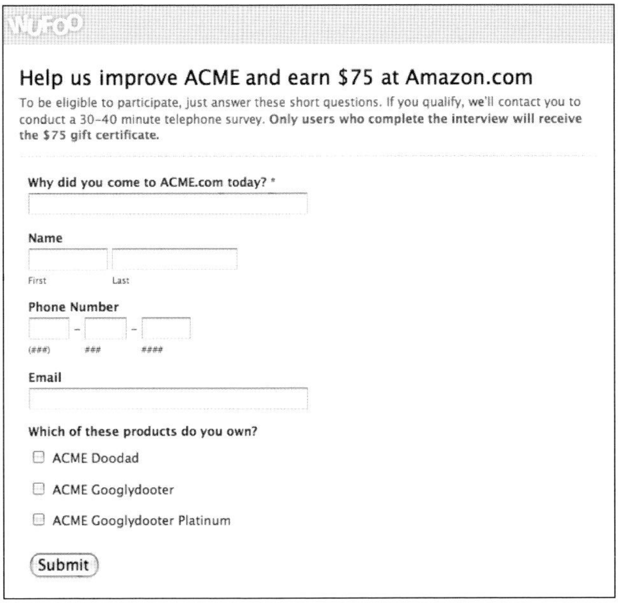

Help us improve ACME and earn $75 at Amazon.com

To be eligible to participate, just answer these short questions. If you qualify, we'll contact you to conduct a 30-40 minute telephone survey. Only users who complete the interview will receive the $75 gift certificate.

Why did you come to ACME.com today? *

Name

First   Last

Phone Number

(###)   ###   ####

Email

Which of these products do you own?

☐ ACME Doodad

☐ ACME Googlydooter

☐ ACME Googlydooter Platinum

(Submit)

FIGURE 3.1
A sample HTML
recruiting form
using Wufoo. You
can embed this
somewhere on your
Web site, or you
can link to it as a
standalone page.

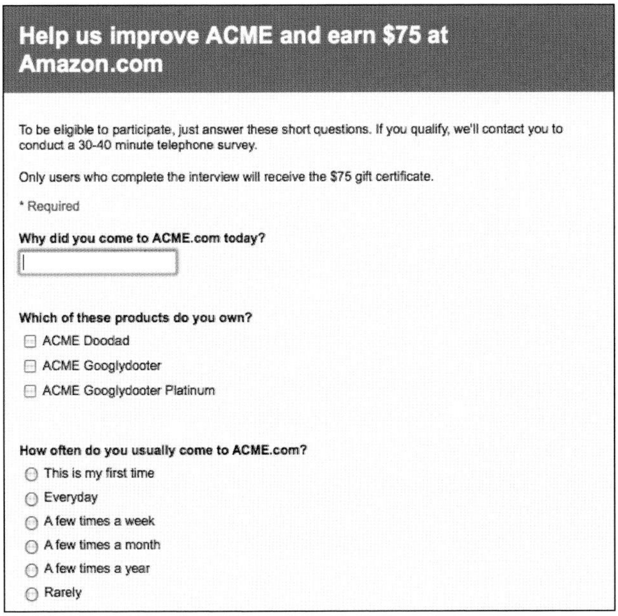

Help us improve ACME and earn $75 at Amazon.com

To be eligible to participate, just answer these short questions. If you qualify, we'll contact you to conduct a 30-40 minute telephone survey.

Only users who complete the interview will receive the $75 gift certificate.

* Required

Why did you come to ACME.com today?

Which of these products do you own?

☐ ACME Doodad

☐ ACME Googlydooter

☐ ACME Googlydooter Platinum

How often do you usually come to ACME.com?

○ This is my first time

○ Everyday

○ A few times a week

○ A few times a month

○ A few times a year

○ Rarely

FIGURE 3.2
A sample HTML
recruiting form using
a Google Docs form,
which is also linkable
and embeddable.

RECRUITING FOR REMOTE STUDIES   53

Contact forms are fairly easy for Web site visitors to overlook. One thing we've learned in our years of UX research is that most visitors often ignore elements on a Web page that don't appear to help them achieve what they came to do. A more effective strategy is to use a pop-up form. While it may be a minor distraction to some of your site's visitors, it is the most reliable way to recruit legitimate visitors to your site for your study, and there are many ways to minimize unnecessary screener display. The most common problem with pop-ups is that most Web browsers nowadays come with pop-up blockers, which prevent your users from even seeing the screener. The solution is to display the pop-up not as a separate browser window, but as a DHTML layer, which is what many pop-up Web ads are.

Logistics of Screener Implementation

To get the code for these recruiting forms and pop-ups onto your site, you may have to jump a few logistical hurdles. If you're in charge of running and maintaining your site, or you work for a small company, you'll need to determine where on the page you want or need to put it. (Ethnio, which we describe later, requires only a single line of JavaScript below the closing HTML body tag. If you don't know in broad terms what JavaScript and a body tag are all about, please tell @boltron on Twitter, and he will gladly recommend some other books for you.)

In larger organizations with more layers of content management protocol and IT bureaucracy, you'll have to arrange things carefully in advance with those who are in charge of maintaining the code. IT operations guys tend to hate external code and will want to know things like what does the code do? Is it secure? What pages does it need to go on? Will it work with our (complicated, Orwellian) CMS? If the IT folks are going to be in charge of implementing the recruiting code, you need to make clear to them exactly what the screener questions should say and how they should appear. In some organizations, changes to the Web site can be made only at certain times of the day or week, so you may need to schedule your study around that.

Higher-ups who are in charge of managing the Web site as a whole will want to know which pages the screener will appear on, as well as how long the recruiting code will be active, what the screener will look like, how many people will be seeing it, if there's a way they can shut it off or disable it on their end, and whether or not the look and feel of the screener will match that of the Web site.

The answers to these questions will vary depending on which recruiting method you're using. For hand-coded forms and pop-ups, you'll need to consult with the people who are implementing the code. If you're using a service like Ethnio, Google Docs' forms feature, or Wufoo, check the FAQs and Help documentation on the respective service's Web site.

Web Traffic Requirements for Live Recruiting

There's a rough formula we follow to predict the recruiting rate for a standard pop-up screener: between 1.5 and 2% of visitors who see the screener will fill it out, and a little over half of those will usually consent to being called. Once you start calling people, usually about 65% of the people you call will be able to successfully participate, depending on your target audience.

If you want to have a steady stream of recruits so you can reliably intercept people right as they come to the site without having to wait long, you'll need **at least six qualified, willing recruits per hour**, which would require site traffic somewhere north of **1,000 visits an hour** during testing hours (see Figure 3.3). In most cases, we find that people with 10,000 daily unique visitors or more do just fine with live recruiting.

You can manage with fewer if you've given yourself plenty of time for testing, but be warned that there may be long periods of time when you're waiting idly for a new and qualified recruit to come in. (Observers should definitely be made aware of this fact ahead of time.)

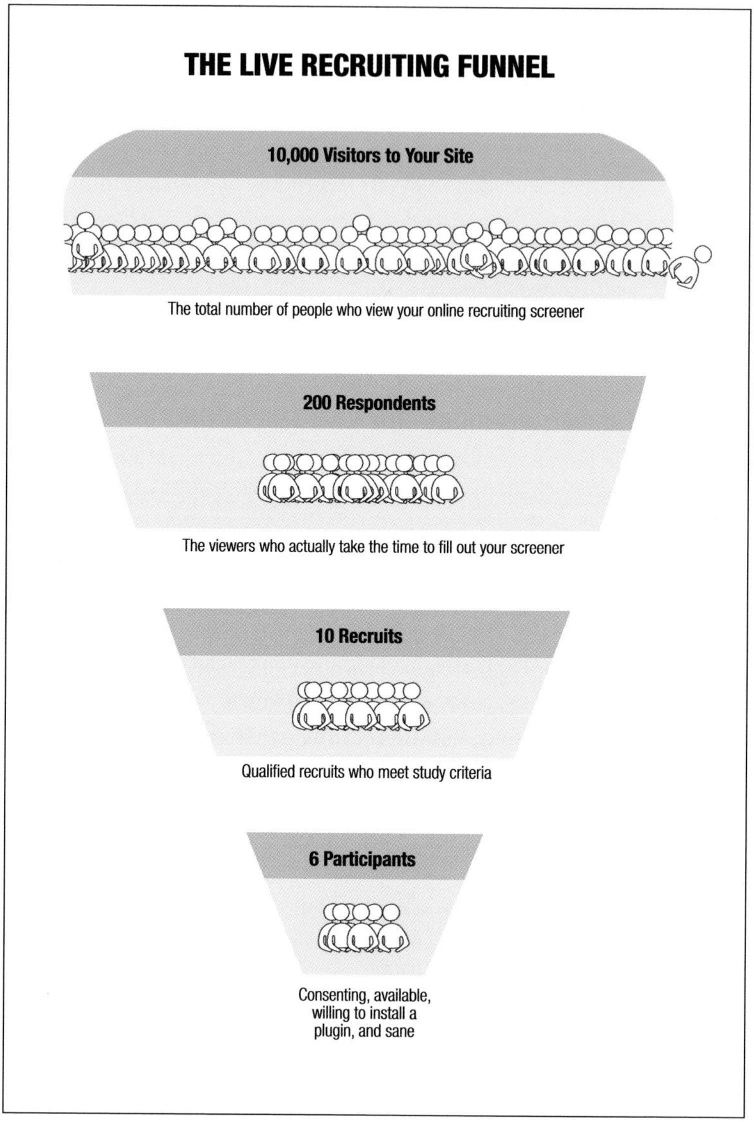

THE LIVE RECRUITING FUNNEL

10,000 Visitors to Your Site

The total number of people who view your online recruiting screener

200 Respondents

The viewers who actually take the time to fill out your screener

10 Recruits

Qualified recruits who meet study criteria

6 Participants

Consenting, available,
willing to install a
plugin, and sane

FIGURE 3.3

The "recruiting funnel": how Web traffic translates to viable recruits. Only a percentage of a percentage of a percentage of your site's visitors will be available for participation.

Ethnio, Our Live Web Recruiting Tool

At the time of this book's writing, there's only one tool or service designed specifically to recruit participants for user research—and as sheer luck would have it, we created it. It's called Ethnio.

When you insert a single line of JavaScript into the code of any Web page (right before the closing body tag: </body>), Ethnio allows you to display a custom DHTML recruiting screener on your Web site. You can badge the screener with your Web site's logo, draft and edit your own questions, adjust the display rate (so that only, say, 50% or 1% of all users see the screener at all), and enable or disable it at any time.

What's this screener like from the perspective of your Web site visitors? Let's say you're visiting ACME.com. As the page loads, you see a pop-up DHTML survey bearing the company's logo, asking if you want to earn (for example) a $75 Amazon gift certificate to participate in a Web site study (see Figure 3.4). After skimming this pop-up, you're curious enough to click it, and you're greeted with a handful of questions, which you fill out and submit in a minute (see Figure 3.5). You get a confirmation informing you that you may be contacted to participate in the study within the hour (see Figure 3.6). You dismiss the screener and carry on with your business on the Web site. Five minutes later, you get a phone call from a moderator, who introduces herself: "Hi, my name's Darlene, and I'm calling about the survey you filled out on ACME.com just a few minutes ago. Do you have time right now to chat with me about the Web site for 30 to 40 minutes?" You say, "Why not?" and the study begins.

FIGURE 3.4

The Ethnio screener's introduction page. This is what some portion of the visitors to your Web site will see.

Ethnio, Our Live Web Recruiting Tool (continued)

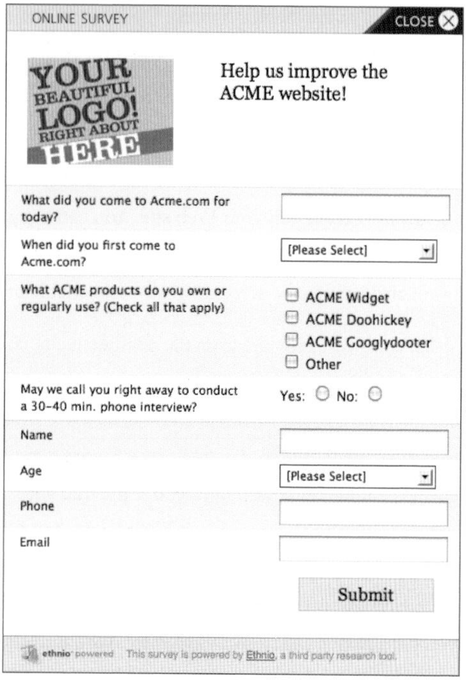

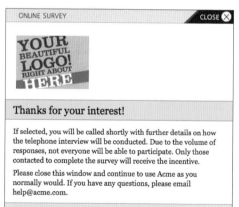

FIGURE 3.5
Ethnio screener questions. You can create open-ended, checkbox, radio button, drop-down, and yes/no question types.

FIGURE 3.6
Ethnio screener confirmation page. Here you usually thank the user for filling out the survey, tell the user when he/she can expect to be contacted, and remind him/her that only participants who are contacted to complete the study will receive the incentive.

Ethnio, Our Live Web Recruiting Tool (continued)

As the visitor's survey response is submitted, it's loaded into a special Web-based Recruits table (see Figure 3.7). The moderator can view and filter this Recruits table in real time and select qualified, consenting recruits as they come in. Later, you can export the Recruits spreadsheet to get a handy snapshot of the responses of everyone who filled out the screener.

Viewing 13 responses of 14 (total)

Add/Edit Filters

Status	Arrival	... today?	... Acme.com?	... apply)	... Interview?	... Name	... Age	... Phone	... Email
New	Thu, Aug 27, 4:26 PM	I wanted to complain about my broken ... googlydooter	Less than a week ago	ACME Googlydooter	yes	amanda lazer	35 to 44	555-555-5555	lazer@beam.com
New	Thu, Aug 27, 4:25 PM	stuff	This is my first visit		no	john nguyen	45 to 54	555-555-5555	ngu@yen.com
New	Thu, Aug 27, 4:24 PM	I love acme products	More than three months ago	ACME Widget, ACME Doohickey, ACME Googlydooter ... Other	yes	sam wise	35 to 44	555-555-5555	samwise123@mordor.com

FIGURE 3.7
The Recruits table. You can watch the responses come in right when the visitors fill out the forms and filter the responses to find the recruits you want.

So how much will this cost you? Prices range from $400 and up, but they also change over time, so go to http://ethnio.com and check it out.

Designing the Screener

The most important step in the Web recruiting process is designing the recruiting screener: the form that Web site visitors will use to submit their recruiting info to participate in the study. A good screener will bring you a steady stream of recruits from which you can choose a representative sample of your target audience.

There are three components to the screener to consider: the introduction, questions, and confirmation.

Screener Introduction

Make the incentive and purpose of the screener immediately clear. Stick with a straightforward, high-visibility header, followed by a very short description:

Help us improve our Web site and earn $75 on Amazon.com

We are looking for visitors to the ACME Web site to participate in a 40-minute one-on-one phone interview about the Web site. To qualify, simply answer these few short questions. If you are selected to participate, you will receive a phone call shortly.

Visitors can easily mistake the screener for a pop-up ad or a sleazy sweepstakes offer, so you'll want to make clear that it's legitimate. A clean design and the Web site's logo (if applicable) can help put people at ease. (This is the biggest visual design challenge—not seeming like an ad.) You also should be clear about how the user will benefit, without making it sound like a promotion or a sweepstakes—no exclamation points, no "You could win X dollars!" Friendly is okay, but you want to avoid sounding gimmicky or excessively jokey.

After the description, either provide a clear link to the screener questions or introduce the questions directly below.

Screener Questions

Screener questions should be unbiased, comprehensive, noninvasive, and strictly relevant to the recruiting goals.

There's only one question you absolutely must ask in all cases, which is the "consent to contact" question: "May we contact you right away to conduct the 40-minute phone interview?" For moderated studies, you'll also want to ask for a phone number to contact the user to begin the session. Aside from these, you should include only questions that are useful for determining whether the respondent is qualified to participate in the study, so keep the question count low. Use as few questions as possible, and no more than 10. Participants tend to abandon long screeners, considering that they may not even get anything for filling it out. The questions should map directly back to the recruiting criteria you defined in the planning stages of the study.

USE DROP-DOWNS EFFECTIVELY

To help cut down on the number of questions you're using, you can often use drop-down menus to address more than one qualification at once. In a study we conducted for Wikipedia, we asked the following: "Have you ever contributed to Wikipedia?" [Yes, started a new article / Yes, edited an entry / Yes, made more than 50 contributions / No, never]. From the response to this question, we were able to determine not only whether respondents had edited entries in the past, but in what way. A cleverly designed drop-down question can answer three or four questions at once, without being too cumbersome.

We like to begin with a simple open-ended question: "Why did you come to the site today?" This question not only helps determine whether a user's motives for coming to the site match the goals of the study, but also helps to root out fakers (an issue we'll go into later). After that, your screener will probably consist of three to five short questions that will help you place respondents in a recruiting category; these questions vary depending on the goals and target audience of your study. When drafting closed-ended questions, make sure that the choices are exhaustive and that each choice unambiguously qualifies or excludes participants based on a particular criterion. For example, if you're selecting participants from a certain age range, be sure all the ages are covered—e.g., "[Under 18 / 18–35 / 36–55 / Over 55]". We advise against relying on too many survey-type questions like Likert scales and exhaustive check box fields for recruiting purposes because respondents generally tend to get bored and rush through them.

If you can, require users to fill out all fields before submitting the form. Disabling the submit button until all fields are filled is a good way to do this.

NOTE **COMMON QUALIFYING SCREENER QUESTIONS**

For most studies, there are a number of pieces of information about users that are nice to know and that can be collected very easily. Age, phone number, email address, gender, location (city/state, not address), size of company, and occupation are sometimes good to know, depending on what kind of interface you're testing. Other common questions include

"When did you become a member of ACME.com?"

"When did you first visit ACME.com?"

"How often do you come to ACME.com?"

"Which ACME products do you own (check all that apply)?"

"How did you find out about ACME.com?"

Remember, you're trying to keep the screener short and focused only on the information that's important for determining participant eligibility. So if you're not using the information to decide whether the user is qualified to participate, ask it as a warm-up question during the actual study session instead. Also, while it's true that you can *often* implicitly infer things like gender or location, don't rely on that inference if knowing that information is important. Just ask.

And now some ethical considerations: you have to assume that some people responding to the screener probably have no idea what will be done with the information they've submitted, so don't abuse that information. On top of the researcher's responsibility to keep basic personal information (names, email addresses, location, etc.) private and secure, you shouldn't request sensitive information: Social Security number, credit card number, passwords, any of that. Recruiting records should be stored, erased, or anonymized in a secure and responsible way. While seeing the users' recruiting info may be useful and helpful, you obviously don't want to face the repercussions if that information gets leaked somehow. This goes double when minors are involved: there are oodles of legal and ethical considerations. You want to make sure that users under age 13 are either restricted from submitting screener responses or that you otherwise follow the appropriate legal protocol in your area. These important consent and privacy issues related to gathering user info are covered in the next chapter. (Also, believe it or not, we belong to the market research [gasp!] association ESOMAR, which publishes strict guidelines for privacy around information gathered in consumer research. You can find these guidelines at **www.esomar.org/index.php/professional-standards-codes-and-guidelines-mutual-rights.html**.)

Screener Submission Confirmation

Respondents often get confused about the screener; some mistakenly think that the screener *is* the study and will occasionally fill your inbox with hostile questions about why they haven't received their $75 gift certificate. So make it clear here that the screener is only to determine *eligibility* for the study and that only people who *complete the interview* will receive the incentive:

> Thank you for your interest! If selected, you will be contacted via telephone within a day with further details on how the telephone interview will be conducted. Due to the volume of responses, not everyone will be able to participate. **Only those contacted to complete the interview will receive the incentive.**

> **NOTE** ALWAYS TEST YOUR SCREENER! ALWAYS!
>
> At least a day before the testing sessions begin, run a test of the recruiting screener. Place the screener on the site exactly as you would if you were actually recruiting, and if possible, get a few people outside your network to fill out the screener using different operating systems and browsers to see whether it's displaying and storing responses properly.
>
> The screener test is crucial for catching any technical glitches ahead of time, as well as for giving you a sense of the volume of responses you can expect to receive during the actual research sessions. If the volume is low, you'll probably need to tweak the screener design (see the "Recruiting Slow?" section later in this chapter).

Paying Remote Recruits

As with any form of user research, you'll probably need to offer an incentive (aka "honorarium") to users to entice them to participate in your study. Unlike in a prerecruited in-person study, however, in a remote research study you won't be able to physically hand a check, gift certificate, or money voucher to your participant after the session is finished, and some people are understandably reluctant to give out their mailing address to a random Internet person who just called them.

For those reasons, we like to offer Amazon.com gift certificates. All you need to be able to send them is the recipient's email address, and we've found them just as effective as cash for purposes of attracting participants. However, if you have something equally appealing to offer that doesn't require users to give out more personally identifying information—discount codes for your company's merchandise, etc.—go for it. If you encounter legal or ethical issues when giving out incentives (some company accountants get skittish about doling out incentives to strangers), you can even offer to donate to a charity in their name.

The amount you should offer to the participants depends heavily on whom you're looking to recruit. The average participation incentive for a typical remote study is somewhere between $50 and $75, which is lower than most in-person studies, since you don't have to incentivize users to travel to a testing location. If you're testing busy high-income professionals (doctors, lawyers, businessmen), you'll need to hike up the amount. We've had to offer as much as $300 to reliably recruit doctors and business executives for a 30–40 minute session. (Be careful: the higher the incentive, the greater the risk of "fake" users. See "Choosing Good Users and Spotting the Fakers" for tips.) On the other hand, you'd be surprised at the number of people who are willing to donate their time for no incentive at all; this is often the case for Web sites with a dedicated community or fan base, or for sites with so much traffic that there's bound to be a few philanthropists in the mix.

Incentives are trickier to handle when testing internationally. Amazon or Amazon-like online vendor gift certificates are still your best bet in most industrialized nations. PayPal can work too. In the developing world, wire transfer of cash might be your next-best bet. Typically, incentives are modest enough that you don't have to worry about running into legal troubles transferring money overseas, but using this type of incentive does mean that you'll have to gather additional information from each participant after each session. The information varies depending on what kind of service you're using, but for most transfers you'll need to determine what the nearest wire transfer location is (e.g., Western Union, American Express), the full name of the person you're wiring the money to, and the city and country where the money will be received. Just make sure that you get all the information you need to fulfill the incentive.

Choosing Good Users and Spotting the Fakers

A few notes on recruiting people from the Internet: the world is what it is. Some people will want to join the study because they have a bone to pick with your company and won't cooperate with anything you ask them to do. Others are professional survey-takers; their responses are slick, canned, and a little *suck-uppy*. ("This site really helps me achieve my goals, and the quality of the design shows me that this is a company that really cares about its users.") And yet others will try to take advantage of your generous offer by fibbing on the screener form to look like an ideal study participant; usually, these "fakers" are terrible participants because they'll be unengaged ("Yeah, it's a good Web site. It's fine. I don't have a problem with it."), or they'll try to say what they think you want to hear—the infamous "Hawthorne effect." These people will not give you good feedback.

The ideal way to deal with these problem recruits is to screen them out from your recruiting responses before you even talk to them. Suspicious patterns and similarities in user details may signal that someone is telling friends about the study. Check the email address domains for groups of uncommon email domains (e.g., @harvard.edu, @dell.com), which may signal that word about your study has gotten out around a particular institution. Since you have the participant's contact phone number, you'll also be able to check for recurring area codes, which often correspond to location. Unless it's important to talk to users from a particular region, it's nice to get a mix of area codes to ease the likelihood that you're talking to a bunch of buddies. This likelihood is even greater if those similar responses are received near one another in time. It could be a coincidence, but be wary.

Another antifaker strategy is to ask at least one open-ended question in your screener and your introductory questions. Most authentic visitors to your site will have a good specific reason for being there, and by asking open-ended questions, you can usually get a strong intuitive feel for authentic visitors. We almost always begin our screeners with the simple question "Why did you come to [Web site name] today?" If the answer is suspiciously vague ("for info," "just looking around," "to see the offerings"), you should be careful to screen the user further if you choose to contact him/her. If the answer is straightforward and specific and fits the study goals nicely ("I came to compare prices between the iDongle and the iDongle Pro"), you can probably be more confident.

On the Internet, info about deals and bargains gets out fast. When your incentive is particularly lucrative or your screener is visible to a lot of people, there's going to be a greater risk of fakers. Be prepared for the possibility that someone is going to see the screener and post about it on some community bargain-hunting or "free stuff" Web site (FatWallet.com, ProBargainHunter.com, Jangle.com, any random message board...). What do you do? Be sneaky and ask questions that trick the fakers into tipping their hand. We like this multiple-choice question: "What's your favorite bargain-hunting Web site?" [Fatwallet, ProBargainHunter, Other, No idea what you're talking about]. If the user selects anything other than the last choice, you'll either want to pass over that recruit or do some extra vetting at the beginning of the session.

Inevitably, however, a few problem users will slip under your radar—see Chapter 5 for tips on how to deal with them.

NOTE   CHECK YOURSELF

The participants are not always at fault when bad participants are selected. Sometimes your own subconscious biases can get the better of your good judgment. Don't cherry-pick users; there's a difference between choosing users who fit the criteria and choosing users just because you think they will be "good participants." The basic guideline to follow to ensure a fair sample when live recruiting is to simply choose the most recent qualified recruit—no picking and choosing on the basis of how good a recruit's spelling is, for example, or just because the recruit irritatingly typed responses in all-caps. That's not to say, however, that you shouldn't dismiss uncooperative users. Again, if a user seems extremely reluctant to talk or appears to have some kind of bad-faith agenda, don't hesitate to politely dismiss that person.

Getting other people who are familiar with the recruiting objectives involved in the recruiting can help mitigate your own biases. Stakeholders, as always, should be encouraged to involve themselves in the recruiting process if they're available so that they can evaluate recruits case by case. It's useful to have many eyes on the incoming responses so that no promising recruit gets overlooked or unfairly neglected.

Recruiting Slow? Don't Sit Around—Fix the Screener

A common problem: it's testing day, and not enough people are responding to your screener for steady recruiting (about six qualified recruits per hour). Your time is valuable, so you shouldn't wait any longer than 10 minutes before taking action to bring in more qualified recruits.

The first thing you should try, if your recruiting solution allows it, is to adjust the screener. Proofread the introductory text. Remember, you want concise, direct prose that doesn't sound promotional. Reducing the number of questions on the screener makes it more likely that visitors will fill out the whole thing. If you're using a screener with an adjustable display rate (i.e., it displays to only a certain percentage of visitors) and the rate is lower than 100%, increase the rate. If this adjustment can be done quickly and easily, consider increasing the display rate to 100% while waiting for a qualified recruit and then bringing it back down once you've got a user on the line. That way, you can get recruits coming in when you need them, and you're not distracting the site's visitors while you're testing.

If you're recruiting from a big company Web site and aren't in charge of making the necessary edits to the page, making these small changes at will is not going to be easy, so the presession screener test well in advance of testing will be even more crucial to get a sense of whether anything needs to be adjusted.

Consider raising your participation incentive incrementally. If you've been offering $50, bring it up to $75 after 15 minutes of slow recruiting, then to $100 after an hour. This approach is always preferable to sitting around and waiting, unless you've scheduled a lot of time for testing and you can afford to sit and wait for the right participant to come in. Remember to get the approval of whoever is financing the incentives. If you raise the incentive *too* high, though, beware of money-seeking fake participants. See "Choosing Good Users and Spotting the Fakers" earlier for more details.

If you have a large Web page and you've placed the screener on a page somewhere deep in the navigation, consider placing it on a more prominent page, like the homepage. You'll definitely get more responses this way. They may not all be as relevant as they were when the screener was on a more specific page, but the overall number of more qualified recruits should be higher.

If you're getting far fewer users than you expect, check to see whether your screener is actually working the way it should. Is it displaying correctly to all users, and do their replies get through to you? You usually catch these glitches during the screener test, but technical bugs, like human feelings, are unpredictable. If you have a suspiciously low number of recruits, make sure that your screener is displaying correctly on different browsers and operating systems. Clear your browser's cache and cookies and try reloading your page again. If you're not tech-savvy enough to make the necessary adjustments, be sure to have an IT guy on hand to help you root out the bugs.

A Professional Recruiter's Advice

by Dana Chisnell

Although we like to think we have this live recruiting thing figured out, we thought it'd be worth getting the opinion of a professional user research recruiter about how to recruit good participants for user research studies.

Dana Chisnell is an independent usability consultant and user researcher who founded UsabilityWorks in San Francisco, California. Since 1982 she has been doing usability research, user interface design, and technical communications consulting and development, conducting user research for companies like Yahoo!, Intuit, AARP, Wells Fargo, E*TRADE, Sun Microsystems, and RLG (now OCLC). She cowrote the *Handbook of Usability Testing, Second Edition*, with Jeff Rubin (Wiley, 2008), and is an STC Fellow and a long-time member of the Usability Professionals' Association and ACM SIGCHI.

A Professional Recruiter's Advice (continued)

THE RECRUITING PROCESS

The sweet spot of timing for most studies, from the time that the selection criteria have been determined to the first session, is about two weeks. If you start any sooner, the respondents' schedules can change, and they might not be able to make it to the study. The first week can be spent on selection criteria; the second week is for recruiting; and week 2 is about sorting, filtering, and completing interviews.

We typically split the screening into two phases. The first phase is prequalifying: "Are you suitable for this study at all?" The second phase tells us whether you're an appropriate participant. From there, we work with the researcher on what the mix of participants should be. Our general approach is to get a checklist of things that the researcher wants to find out about for the study, focusing on behaviors, and then based on those, coming up with a set of open-ended questions. We encourage researchers to look at recruiting as an important part of the research they're doing.

USING OPEN-ENDED QUESTIONS TO MODIFY GOALS AND CATCH FAKERS

By asking open-ended questions, we learn things that may change the user profile we're looking for. For example, in one study we thought we wanted only people who booked their own travel, but then we saw that many of the respondents didn't book their own travel—an assistant or spouse did it for them. So we realized we wanted to talk to those people instead.

Sometimes fakers can get past a screener because they can infer what the study's about from the closed-ended questions they're given, as well as what kind of user the recruiter is looking for. So when we talk to respondents over the phone, we always include open-ended questions. For example, when we recruited for a video game study, we asked respondents who their favorite characters were in the game and why. That requires a deeper knowledge about the game than what someone can read on the back of the box at the store or look up online, and if they don't know the game, then they wouldn't answer that question well. Sometimes respondents ask you a lot of questions in return because they're trying to get all the info they need to answer the question.

Some "cheaters and repeaters" will show up on your radar a lot for different studies because they're always responding to study invitations. That's not to say that everyone who responds to multiple calls for participants over a year is a faker. These people may actually qualify for many different studies. But typically we don't want career test participants.

A Professional Recruiter's Advice (continued)

DISMISSING RESPONDENTS GRACEFULLY

We usually go through the whole screener even if the respondent isn't going to qualify for the study. Throughout the process, we remind respondents that they're not necessarily eligible for the study yet, and at the end of the interview, we make it clear that this was a qualifying step. We'll say, "Thanks a lot; now we'll talk about your availability." If a user's qualifications are in doubt, we'll say, "We're still fitting people into the right mix for the entire study, and we'll call you back and let you know if you're eligible to participate." We always call users back to let them know one way or the other. Sometimes, if we discover that a user isn't right for the study, we'll say, "You know what, we've filled all the appointments for the particular group that you fit into; we really appreciate your time, and if it's okay with you, we'd like to hang on to your contact info in case we have an opening in a future study. Would that be okay?"

We include a standard paragraph in our recruiting call and forms that says, "We will do our best, but we may not be able to contact everyone. If we contact you, we may ask additional questions before we make selections. You might not be selected for the study."

VETTING RECRUITS

You don't want participants who are evasive, don't want to spend time on the phone with you, are reluctant to commit to any particular time, aren't responsive to emails or phone messages, or are fluffy about whether or not they're going to show up.

As for closed-ended questions, it's good to provide ranges that you can use to sort out users. If you're using multiple-choice questions asking about experience with a type of product, offer a list of products to make it hard to guess which or how many you're most interested in. If you can randomize the order of the choices, you should. If you're looking for a certain level of experience with particular products, it's also useful to use frequency matrices, where you can list certain behaviors or activities along one axis, and frequency along another; For example, you could ask how often people engage in certain behaviors online, and they could respond, "Daily," "Weekly," "Monthly," and so on. And then you can come up with a rubric to calculate the desirability of the participant on the back end.

Wait! Read On...

Before you start recruiting real users from your Web site, read through
the next chapter to make sure you have all your legal privacy and consent
obligations sorted out. It's very important!

Chapter Summary

- Live recruiting is a method of getting people who visit your Web site to participate in your remote research study.

- Live recruiting is what makes Time-Aware Research viable, because you can catch people just as they're about to perform a task you're interested in observing.

- The most reliable way to live recruit is by placing a form or a DHTML pop-up (the "recruiting screener") on your Web site, using services like Wufoo, Google Docs, or Ethnio.

- For larger organizations, there will be organizational obstacles for placing the recruiting screener; make sure that your IT department and upper management are on the same page as you.

- You'll probably need at least 1,000 unique site visitors per hour for viable live recruiting.

- Keep your recruiting screener short and to the point: no more than 10 questions, including the obligatory consent-to-be-contacted question.

- In lieu of cash, online retail gift certificates are effective, easy-to-fulfill participation incentives; the gift amount depends on how hard the participants are to find. Merchandise and even volunteer requests can also work sometimes.

- Some people will try to cheat or game the recruiting screener. Scrutinize the responses carefully, especially open-ended ones, to spot fakers. Sometimes an anti-faker trick question will be necessary.

- If you're not getting enough qualified recruits in spite of decent Web traffic, proactively edit your recruiting screener and incentive ASAP, instead of wasting time waiting for the right users.

CHAPTER 4

Privacy and Consent

In all forms of research, obtaining consent from participants is the part that exposes the researcher to legal liability, which means it's definitely something you need to pay attention to. With remote research and recruiting from the Web, there are a number of laws surrounding contacting, observing, recording, and collecting information from people online and over the telephone. Don't let these laws intimidate you, though. We'll address some methods to cover your bases for a typical study in the United States, provide advice for the more complex world of international research, and tell you everything you need to know to form your own rigorous approach to legal consent in remote testing.

Certain Assumptions

We'll assume that you're using all the info collected (both from recruiting and from the research) strictly for internal behavioral user research and that you'll remain well within reasonable ethical boundaries of information gathering for those purposes—that you're not doing evil things like collecting Social Security numbers, selling participant info to other companies, using participant quotes and recordings in your marketing, soliciting info that might be compromising, harmful, or humiliating to the participants, subjecting the participants to unethical or uncomfortable tasks, etc. We can't help you there. We'll also assume that you follow the best industry practices in securing your data—keeping it off any public networks and keeping any consumer information secure and password protected.

The first few sections of this chapter cover most typical studies; the later sections deal with the daunting special cases of international studies and studies involving minors.

Last important thing: even though we discuss some legal matters in this chapter and consulted informally with legal professionals while writing this chapter, we're no legal experts ourselves, and as such **we're not claiming to dispense legal advice** in any way, shape, or form. What we're doing is describing a general approach that we've been comfortable with in the past for our own specific studies. It's impossible to generalize about legal matters for every type of study, and the only 100% foolproof way to know your personal privacy and consent issues is to talk to a lawyer specializing in Internet law. Just saying.

Set Up Your Web Site's Privacy Policy

If you're going to be recruiting from your Web site, you need to make sure that your Web site (or the service that's in charge of gathering the user data for recruiting) has a privacy policy. That's not to say that every recruit has to read through it; this information just has to be somewhere fairly easily visible and accessible on the Web site. Such policies are pretty standard, as most Web sites that allow for user registration already have them.

Your privacy policy needs to tell your Web site's visitors a number of things. It needs to detail exactly what info is being collected from recruits, what you do with the info, and how recruits can change the info, including any related means to contact the Web site administrators. If the way in which you're using the visitors' information changes at all from the time you collected their information, you need to be able to contact the recruits, informing them of the update to the privacy policy. The policy should also include your organization's legal name, physical address, contact information, a statement indicating whether you're for-profit or nonprofit, and the name and title of an authorized representative.

If your Web site doesn't have a privacy policy that will enable you to recruit users, we highly recommend retaining the services of an attorney to draft one for you. A competent Internet law expert can whip up a simple one in an hour. (Software and online services do allow you to create a policy yourself, but this is a one-time expense, so why risk botching it?) You can find a generic sample privacy policy at the Better Business Bureau Online (BBBOnline) Web site at www.bbbonline.org/privacy/sample_privacy.asp. There are tons of additional rules if you're going to be collecting any information from minors under age 13, and we'll discuss that topic in "Consent for Minors," later in this chapter.

Basic Participation Consent

As with any study, you need explicit consent from recruits to participate, especially if you're going to be recording the session. You'll really need consent if any of the information or data collected from users is sensitive in any way. And you'll super-royale-with-cheese need it if you're testing minors or international users. A *consent agreement* is a contract that makes the terms of the study clear (what participants will be required to do for the study and how they'll be observed and recorded) and then obtains explicit agreement to those terms.

Obtaining consent for a remote study is not too different from obtaining consent for an in-person study. For studies within the United States, all you'll need is a consent agreement statement, signed with some form of tangible and unambiguous consent that can be identified with the participants. It can come in the form of a signature, an explicit verbal agreement, or a *clickwrap agreement*, which is a form response that users submit to signal that they're willing to agree to the terms of the study. (You're probably familiar with the Terms of Service/EULA agreements that come up when you're installing software or registering for Web services; those are examples of clickwrap agreements.) Since getting someone's signature over the phone is not very easy, you'll probably want to use a clickwrap to get a legal record of your participant's consent. Clickwrap agreements have been held enforceable in court a number of times in the past and should provide enough legal traction to enable you to call users to participate in remote studies.

If you're using the live recruiting methods we explained in the preceding chapter, it just so happens that you already have a viable clickwrap form: the recruiting screener. As we mentioned in Chapter 3, always include a clear Yes/No question that asks users if you may contact them to participate in the study. Again, the question we use is this:

> "May we contact you right away to conduct a 40-minute phone interview?"

In the 10 years we've been doing this kind of research, we've never run into any problems using this phrase. It's clear, unambiguous, and provides an easy way for users to signal their intent to consent. This question should have a required response, and the form entry should not default to having the "Yes" option selected.

A clickwrap won't necessarily suffice for *any* agreement you want to put into the screener. You couldn't, for example, ask people to consent to something ridiculously unreasonable like "I agree to waive all my rights to privacy during the session" and expect a clickwrap to cover you. However, it should be fine for gathering consent to be called to participate in a remote behavioral research study.

> **NOTE** WHEN CONSENT IS INVALID
>
> Not everyone is legally capable of agreeing to a contract. You should be particularly concerned when the person signing the contract is
>
> - **A child** under the age of the jurisdiction required for a contract to be enforceable. This age varies from state to state in the United States; it's usually at least as low as the legal age of marriage. (Duh.) For user research purposes, a parent or guardian can give permission for a child to participate, although this can get really messy for children under 13 (see "Consent for Minors.").
>
> - **Mentally impaired**.
>
> - **On medication** or medical care that makes it difficult to make clear, fully informed choices (interviews after surgery, or injury, or while on medications that influence thinking).
>
> - **Under considerable stress**, making it more difficult to make clear, fully informed choices (interviews right after a disaster, for example).
>
> - **Very old**, to the point at which it's reasonable to question whether consent is completely understood.
>
> - **Not fully conversant in the language** of the consent agreement.

Generally, a contract signed by someone without legal capacity is null and void, as if it never existed, so that even if you have a signed contract, it's as if consent was never actually obtained. So make sure your participants are legally capable!

Children under 13 should already be screened out (see "Consent for Minors" later in the chapter), and in the other cases, you should be attentive to any communication difficulties during the introductory/warm-up phase of the session. If the recruit doesn't seem reasonably lucid or responsive, it's best to err on the side of caution and find another participant.

For many run-of-the-mill studies, verbal consent to participate should suffice; the users' participation as shown in the recording should make clear their willingness to participate. Of course, sometimes you'll want more heft in your consent form, spelling out all the terms in advance and gathering more information from the users that you can use to make their consent really ironclad. This is the case for any study in which the liability risk is higher than normal or in which you expect to be exposed to more personal information than usual—e.g., studies involving medical, financial, or government-related info; info about family and other individuals; religious or political info or opinions; or any info that could reasonably be considered "intimate."

For these studies, you'll want to make details about the study explicit to the users in a consent agreement statement (see Figure 4.1). You can include this statement in the body of your screener, either embedding it or linking to it on a separate page. Or if you want to keep the screener brief, you can also wait until you've contacted participants before you direct them to the full consent statement. You can simply follow the usual recruiting process to contact users, and after you've got them on the phone and have asked them whether they'd be willing to participate in the study, direct them to a page with a form that lays out the details of the study in plain, unambiguous terms. (See the following text for examples.)

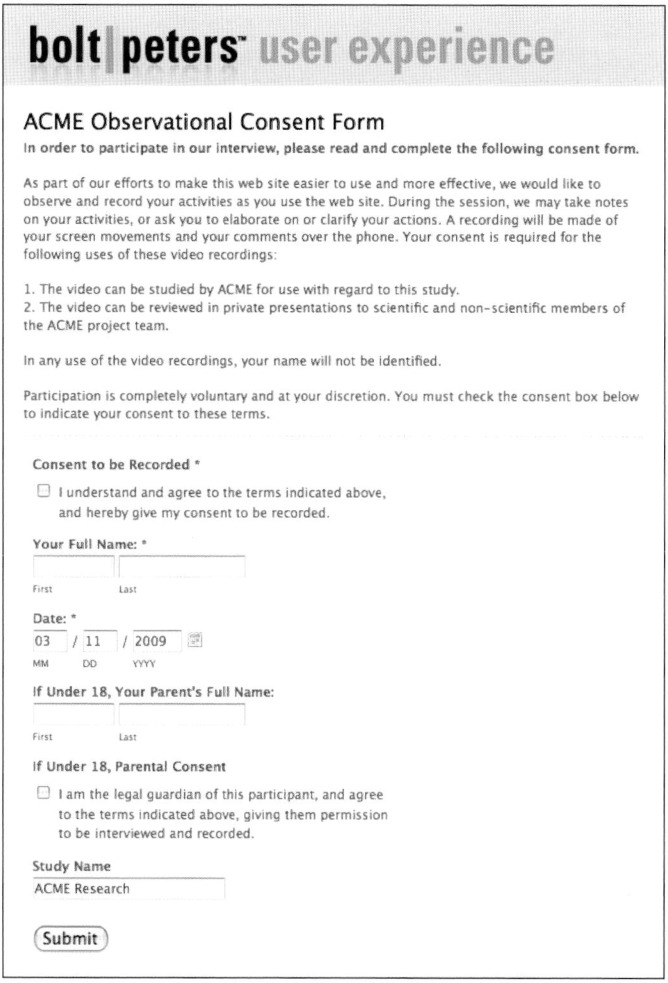

bolt|peters user experience

ACME Observational Consent Form
In order to participate in our interview, please read and complete the following consent form.

As part of our efforts to make this web site easier to use and more effective, we would like to observe and record your activities as you use the web site. During the session, we may take notes on your activities, or ask you to elaborate on or clarify your actions. A recording will be made of your screen movements and your comments over the phone. Your consent is required for the following uses of these video recordings:

1. The video can be studied by ACME for use with regard to this study.
2. The video can be reviewed in private presentations to scientific and non-scientific members of the ACME project team.

In any use of the video recordings, your name will not be identified.

Participation is completely voluntary and at your discretion. You must check the consent box below to indicate your consent to these terms.

Consent to be Recorded *

☐ I understand and agree to the terms indicated above, and hereby give my consent to be recorded.

Your Full Name: *

| First | Last |

Date: *

| 03 | / | 11 | / | 2009 |
| MM | | DD | | YYYY |

If Under 18, Your Parent's Full Name:

| First | Last |

If Under 18, Parental Consent

☐ I am the legal guardian of this participant, and agree to the terms indicated above, giving them permission to be interviewed and recorded.

Study Name
ACME Research

(Submit)

FIGURE 4.1
A sample Wufoo-generated extra-strength observational consent form. You can direct the participant to this form either by sending a link or by reading the Web address aloud.

Sample Consent Agreement

Thank you for participating in our study. This Consent Agreement is a legal document. Please review it carefully and let us know whether you have any questions. You are free to discuss this consent agreement with anyone that you choose.

Who We Are. We are ACME Inc., a California company that produces widgets. We would like to learn more about how people interact with our Web site so that we can improve our business consulting services.

Recording and Observation. We will observe your interactions with the ACME Web site and will take notes about your interactions with the Web site. We will make a video and audio recording that will include your voice, as well as your computer screen, only for the duration of the session. We expect this will take about 40 minutes of your time.

Our Use of Notes and Recording. We will use our notes and the recordings internally within the company. We will not distribute copies of these recordings to persons outside our company—for example, we will not sell copies or post excerpts on our Web site, or use the recordings in advertising. We will not identify you publicly as a participant in our project. The recordings will be digitally stored on our secure servers. We will give you a copy of our recording of you if you request it in writing.

Compensation. Compensation for completing your participation is a $75 Amazon.com gift certificate, which will be delivered to your email address 1–2 weeks after the session.

Contact Us! We are happy to answer any questions that you may have, so please do not hesitate to contact us. We can be reached by phone at (XXX) XXX-XXXX or by email at research_moderator@acme.com. Please let us know if you want a copy of this Consent Agreement.

Consent to participate

❏ I understand and agree to the above terms, and give my consent to participate in the research study.

[Submit Consent Agreement]

The Blue Jeans Debacle

In the early years of remote testing, nobody had done live recruiting before; we hardly knew all the things that could go wrong with all this info. One case in particular, which we now remember wistfully as "The Blue Jeans Debacle," illustrates the kinds of things that can go wrong when you're not careful with user data.

Before pop-up blockers became widespread, we had to hand-code PHP pop-up screeners on a project-by-project basis to recruit our users. Naturally, it was our policy to store all user-submitted data in secure, protected databases, but for one retail client, we had a bug in the code that made some of the user data visible on Google in a plain-text file. It disclosed people's first and last names, along with their preferences for buying blue jeans. In a matter of hours, we received a handful of angry emails from people who had happened across this information after googling their own names. They wanted answers.

We had to do quick damage control. We disabled the recruiting screener, sent out apologies to the affected users, and fixed the bug in the code. It was our stupid good luck that the leaked info wasn't vitally sensitive, but of course no users want any of their personal information shared without their knowledge. The people we contacted were very gracious after we'd done our explaining, but the situation could have been worse.

This little anecdote demonstrates two points. First, as in any case in which you're collecting user data, you must be extremely careful that it's being stored securely and properly. Test your screeners before recruiting so that you know where the data is going. Second, it's best to require the bare minimum of sensitive information you need to do useful research, because if that info does ever end up getting out, it's better that it's a preference for blue jeans and not, say, a home address. Don't ask for any sensitive info in your screener; this information should be gathered during the research sessions.

When you're drafting your own consent statement, there are a few stylistic guidelines you can follow. Clarity is key. Short sentences and paragraphs are easier to read than long blocks of text. Organize the sentences and paragraphs to lead readers through the issues, using headings if the statement gets longer than 250 words. Use simple language. It has to be understandable at an eighth-grade reading level. Don't use words like "aforesaid," "said" (as in "said recordings"), "hereinafter referred to as," "hereunder," "thereunder," "witnesseth," or "for good and valuable consideration, receipt of which is hereby acknowledged." Those words are all confusing and vague; plus, they make you look pretentious and lame.

Consent to Record

In the United States, there are federal and state laws regarding the legality of recording telephone conversations.

> **NOTE** ONE-PARTY AND TWO-PARTY STATES
>
> In the United States, there is a federal law that states that in a telephone conversation between two or more parties, at least one of the parties must consent to be recorded in order to legally record the conversation.
>
> Beyond that, there are individual state laws as well. States can be either "one-party" or "two-party" states. One-party states require the consent of only one of the parties in the conversation to be recorded, whereas two-party states require the consent of *everybody* who's in the conversation. When you call across state lines, if either side of the conversation is in a two-party state, you need to abide by the laws of the two-party state. The two-party states are Connecticut, Florida, Illinois, Maryland, Massachusetts, Montana, Nevada, New Hampshire, Pennsylvania, and Washington. In Michigan, anyone who's participating in the conversation can record the conversation, but any outside party listening in on the conversation needs to get the consent of all participating parties. All other states are one-party states.
>
> For more info on telephone recording, consent, and wiretapping, check out the Reporters Committee for Freedom of the Press Web site at **www.rcfp.org/taping**.

If you're recording the session in any way, you should get explicit consent for all the different recordings you're doing. We're not aware of any laws specifically regarding the recording of other people's computer screens and webcam feeds, but it's just good common sense to get at least the same level of explicit consent that you're getting for the telephone recordings.

The Federal Communications Commission (FCC) describes three ways to get permission to record a telephone conversation:

1. You get verbal or written consent to record from all parties before the conversation begins;

    or

2. You give the participant verbal notification about the recording, which is recorded at the beginning as part of the call, by whoever is recording (i.e., you);

    or

3. You play a "beep tone" sound at regular intervals over the course of the conversation, indicating that the session is being recorded.

For the sake of thoroughness, we recommend *both* of the first two options, asking for permission to record both before and immediately after the recording has begun. We like to start the recording by saying

> "Okay, so now the recording has begun, and I'm able to see your screen. Can I reconfirm with you that you agree to participate in this remote research study, and that you understand that both the conversation and your computer desktop will be recorded for research purposes, until the end of the session? The recording won't be used for anything other than our own research, and we won't share it with anyone else."

Once you've jumped through these hoops, you're free to begin testing.

NOTE CONSENT HELPS BUILD TRUST

All this fuss about consent isn't just to cover your own ham hocks; it's also there to give your participants some peace of mind about your trustworthiness. At this point in time, most people aren't used to being called right away after filling out forms on the Internet, so when you get consent from users, you're also helping to assure them that you're not going to take over their computers, abuse their information, or do anything else underhanded.

At any rate, you'll still be amazed that most users are totally willing to participate in these kinds of studies without hesitation. Since they've already filled out the screener that prepares them to be contacted, it's not as if you're cold-calling them. We've found they're usually pleasantly surprised at being contacted.

International Consent

Testing internationally is trickier for two main reasons: language differences can make it harder to gather informed consent, and consent laws vary for every single country. Covering every single case for every single country for every set of research goals wouldn't just take a whole book; it would fill a whole law library. In the following sections, we'll touch on general approaches for getting participants in other countries, pointing out the considerations where it's important to have due diligence in your consent gathering.

Recruiting Services

The first thing we'll acknowledge about international consent is that it can be easier to obtain consent by recruiting through other methods than do-it-yourself live recruiting. The simplest option is to hire a recruiting agency or research facility to find the participants. There are international services that do this kind of recruiting, and also companies native to the country you're testing in. We've gone through Apogee Group Recruitment to find users all the way over in Hong Kong, Acorn Marketing and Research Consultants in Singapore, and Cooper-Symons Associates in Australia. (You can probably find what you're looking for by googling "research participant recruiting agency [city/country name].")

Many research services will also assume the responsibility for obtaining participation consent and arranging incentives so that you don't have to deal with international money transfers yourself. That's a big plus.

Hiring a Local Lawyer

Another way to approach recruiting is to contact a qualified lawyer in your testing region, preferably one who specializes in privacy and consent law, to draft an international consent form in the native language, tailored to the goals of your study, which you can place online for recruits to fill out. We're told that drafting a consent form for a typical UX study would require a few hours of work.

Here are some bits of info you should be prepared to discuss with the lawyer:

- Your research objectives.

- What kinds of information will be collected (especially recordings and personal info).

- How the information will be used, applied, and stored.

- Who will have access to the information.

- Who will be contacted to participate, and how they will be contacted.

- What participants will be expected to do during the research session.

- How participants will be compensated.

- How participants will be able to contact you with any questions afterward.

Local Research Practitioners

It never hurts to reach out to other research practitioners in the country you're testing in to ask them what they think the major consent and privacy issues of the region are, or whether they know of any useful resources or legal experts to refer to if you want to learn more. UX practitioners can be found working as consultants, working for large companies, or on professional networks like LinkedIn. Don't be shy about contacting market researchers and academic researchers too, since they encounter many of the same issues. As always, use due diligence to determine whether the practitioners are reputable.

Note also that, though local practitioners may know plenty about testing *within* their country, they may not be aware of the issues surrounding testing *between* countries (and there are usually plenty). So this is more of an extra step than a complete strategy for learning everything you need to know.

International Consent Example: EU's "Safe Harbor"
[Note: We've adapted most of this info about Safe Harbor laws into lay-speak from the Department of Commerce's Web site, at http://export.gov/safeharbor.*]*

If you plan on striking out on your own, here's an example of the kinds of rules and regulations you can expect to learn about. The U.S. Department of Commerce maintains a set of principles called "Safe Harbor," dictating the privacy standards all U.S. organizations must comply with when dealing with people in the European Union. You need to adhere to the seven Safe Harbor principles and then certify your adherence.

The seven principles are summarized as follows:

- **Notice.** You need to let your recruits know about the purposes for which you're collecting and using their information, providing them with information about how they can contact you with inquiries or complaints, any third parties to which you'll be disclosing the information, and all the choices and means you'll be offering the recruits for limiting the use and disclosure of their info.

- **Choice.** Recruits must be given the opportunity to "opt out" of having their information disclosed to a third party or used in some manner other than the one for which it was originally collected or subsequently authorized by these individuals (i.e., screening and contacting the recruits for the study). When sensitive information is involved, you need the users' explicit "opt-in" consent if you're going to disclose their info to a third party or use it for something other than screening or contacting for the study.

- **Onward Transfer.** To disclose information to a third party, first you have to apply the Notice and Choice principles. You can transfer information to a third party only if you make sure that the third party subscribes to the Safe Harbor principles or is subject to the Directive or another adequacy finding. Another option is to enter into a written agreement with the third party, requiring that it provide at least the same level of privacy protection for the recruits' information as the Safe Harbor principles require.

- **Access.** All recruits have to be able to access the personal information you've gathered about them and then be able to correct, amend, or delete that information where it's inaccurate, unless the burden or expense of providing this access would be disproportionate to the risks to these individuals' privacy or where the rights of persons other than the recruits would be violated.

- **Security.** You have to take "reasonable" precautions to protect personal information from loss, misuse and unauthorized access, disclosure, alteration, and destruction.

- **Data Integrity.** The personal information you collect has to be relevant for the purposes for which it is to be used, which in the case of recruiting means that all the screener questions need to relate to your study goals. You should take "reasonable" steps to ensure that the data you collect is reliable for its intended use and also that it is accurate, complete, and current.

- **Enforcement.** You'll need to have three things to make the Safe Harbor principles enforceable. First, you need readily available and affordable mechanisms that allow individuals to file complaints and have damages awarded. Second, you need procedures for verifying that the commitments companies make to adhere to the Safe Harbor principles have been implemented. Finally, you need to have a plan to remedy problems arising out of any failure to comply with the principles. The sanctions must be rigorous enough to ensure your compliance. You need to provide annual self-certification letters to the Department of Commerce to be covered by Safe Harbor.

To certify your adherence to the Safe Harbor principles, you can either join a self-regulatory privacy program that adheres to the Safe Harbor requirements (BBB OnLine, TRUSTe, AICPA WebTrust, etc.), or you can self-certify.

To self-certify, you'll have to submit a four-page Safe Harbor application form (which can be found on the Web site) to the Department of Commerce, along with a $200 registration fee. As long as you want to be covered by Safe Harbor, you'll also need to recertify every year, which costs $100.

And that's just the basics. Kinda complicated, huh? But that's the kind of stuff you'll have to brush up on if you're going to brave the waters of international testing for the first time, all by yourself. Nobody said it'd be easy. Once again, complete details about Safe Harbor can be found at http://export.gov/safeharbor.

> **NOTE**   WHAT'S MY LIABILITY?
>
> Failure to disclose the terms of your information gathering and usage in sufficient detail may expose you to claims of fraud, deception, invasion of privacy, and intentional infliction of emotional distress.
>
> Breaking privacy or consent laws can subject you to really breathtaking fines from the Federal Trade Commission (FTC), Department of Commerce, or other government agencies. On top of that, if you're using the information or consent for illegal purposes (spying on users, selling off personal info, etc.), you can get fined far worse than that.
>
> For broadcasting without informing the people being recorded, the FCC can fine you up to $27,500 for a single offense and no more than $300,000 for continuing violations.
>
> And violating COPPA (see the following section), probably the touchiest of all these legal concerns, can cost tens of thousands of dollars in the minor cases, and up to $1 million in the most serious case yet (Xanga in 2006).
>
> Please stick to the law.

Consent for Minors

At last, we've reached the infamous liability snake pit: minors. In the United States, there's a law called the Children's Online Privacy Protection Act of 1998 (COPPA), which lays out the requirements for Web sites that collect personal information from children. So first, the good news: COPPA covers only children under 13 years of age, so testing teenagers (age 13–17) isn't much different than usual. Still, we urge you to take precautions with teenagers, less for the sake of liability and more because you don't want parents to get anxious about their children talking to strangers on the phone. Make an extra-strength clickwrap agreement for both the participant and the participant's parent to fill out. The participant's parent should consent to having his/her child participate in a 30–40 minute research study, with the understanding that the child will be observed by the moderator with screen sharing software. You should also encourage the parent to observe the session if he/she wishes.

So, then, what to do about minors under 13? Our legal guy said this: "Every single time I've had a client who was starting a Web site and wanted to contact children under the age of 13, 100% of the time I have convinced them not to do it. From an administrative standpoint, it's really, really painful." There you have it, gentle readers. Hiring a recruiting agency to deal with contacting minors can lift the tremendous burden of contacting, recruiting, screening, and gathering informed consent from minors and their parents.

But, okay, what *if* you wanted to do your own recruiting, for whatever reason? We'll tell you one thing: you probably won't be able to do "live recruiting" in the sense that you can intercept the user in the middle of a natural task. Gathering the proper consent will most likely take a considerable amount of interruption, since it involves getting a lot of heavy-duty parental identification and consent. Recruiting over the Web with a screener, however, is still an option; you'll just have to switch up your approach. A clickwrap isn't gonna cut it here, and neither is an email from the parents' email address telling you that they consent. To comply with COPPA, you need rock-solid "verifiable parental consent" (see the sidebar). You shouldn't even bother targeting the minors directly. Instead, your screener should target parents who may be willing to allow their children to participate in a remote study, or else it should direct any minors who might see the form to get their parents to fill it out for them.

In short, recruiting minors under 13 is a colossal migraine, a lot of work, and incredibly liability prone (fines begin in the tens of thousands and go up past the million dollar mark). If you don't have the budget for a recruiting agency, you're bound to follow COPPA rules. See the FTC's "How to Comply with The Children's Online Privacy Protection Rule" (www.ftc.gov/bcp/edu/pubs/business/idtheft/bus45.shtm).

What's Verifiable Parental Consent?

Verifiable parental consent is a really intense way of affirming (1) that the parent consents for his/her child to participate in the remote study, and (2) that the parent really is who he/she claims to be. To have verifiable parental consent, you'll need to do two things: give the parent a form that explains the terms of the study and all the information gathered from the child, which he/she can then sign, date, and mail/fax/scan back to you, and then verify the parent's identity.

So first, you'll need the parent to sign a hard-copy informed consent form and fax or send it to you, along with a photocopy of his/her government-issued ID (driver's license, passport, etc.). Then you'll have to verify that the parent is actually the parent. You can do this any one of the following ways:

- Provide a consent form for the parent to print, fill out, sign, and mail or fax back to you, aka the "print-and-send" method; OR

- Require the parent to use a credit card in connection with some transaction (yes, this means you'd have to charge for something); OR

- Maintain a toll-free telephone number staffed by trained personnel for parents to call in their consent; OR

- Obtain an email with the parent's consent, which also contains a digital signature.

If you're not going to be sharing the information with any third party, you have two other "email plus" options, which involve just emailing the parent to ask for consent and having him/her reply with consent:

- In your initial email seeking consent, ask the parent to include a phone or fax number or mailing address in his/her reply so that you can follow up to confirm consent via telephone, fax, or postal mail; OR

- After a reasonable time delay, send another email to the parent to reconfirm his/her consent. This confirmation email should have everything from your first email and also tell the parent that he/she can revoke the consent and inform him/her how to do that.

Privacy Policy for Minors

Most generic privacy policies state that the Web site doesn't collect information from minors under 13. If you are collecting such information, though, you have to make a bunch of amendments to your privacy policy. The complete set of rules, "Drafting a COPPA Compliant Privacy Policy," is at www.ftc.gov/coppa. The gist of it is that your Web site has to do six things:

- Link prominently to a privacy policy on the homepage of the Web site and from wherever personal information is collected.

- Explain the site's information collection practices to parents and get verifiable parental consent before collecting personal information from children (with a few exceptions).

- Give parents the choice to consent to the collection and use of a child's personal information for internal use by the Web site and then allow them to opt out of having the information disclosed to third parties.

- Provide parents with access to their child's information and the opportunity to delete the information and opt out of the future collection or use of the information.

- Not make a child's participation in the study require disclosing more personal information than is reasonably necessary for the activity.

- Maintain the confidentiality, security, and integrity of the personal information collected from children.

As before, we strongly recommend getting a legal expert to either draft such a policy or to look over your draft to make sure everything checks out.

Don't Record Minors

Not even verifiable parental consent will necessarily allow you to record minors. In some states there have been cases in which the parents' consent for the minors to be recorded on behalf of the minors—"vicarious consent"—hasn't held up in court (like *Williams v. Williams* in Michigan, 1998), so you just don't want to risk it. We advise capturing your sessions the old-fashioned way: take really good notes.

Chapter Summary

- For a remote study, you need to obtain various forms of consent for contacting recruits, getting people to participate, and recording the session. There are important additional requirements for international users and minors.

- In order to collect any data from Web site visitors, your site needs to have a Privacy Policy.

- You can obtain participation consent for a typical domestic nonminor user study by using an online Consent Agreement clickwrap form. Make sure the user can give valid consent and that you're not collecting sensitive information.

- We recommend obtaining consent to record the session by confirming verbal consent both before and after the recording has begun.

- Learning the ins and outs of international consent is simpler if you consult recruiting agencies, lawyers, and user researchers who work in the region you're testing in.

- When testing minors aged 13–17, obtain an extra-strength clickwrap consent from both the participants and their legal guardians. Testing minors under 13 is tremendously inconvenient and requires verifiable parental consent and compliance to COPPA laws. Don't record minors.

CHAPTER 5

Moderating

Moderated research is about talking to people. No matter how well you've memorized the study objectives, set up the equipment, and recruited the right users, that doesn't mean squat if you don't establish good working communication with your participants.

Skeptical UX researchers accustomed to in-person research often ask us how it's possible to effectively establish rapport with users without being physically present. Conventional wisdom dictates that you can read more from in-person visual cues, create a more comfortable and natural conversation, and make better motivational inferences when you're face to face, but the truth is, for much of technology research, people are perfectly comfortable talking on the phone while in their own computing and physical environments. (Since moderating with only the phone is such a hot topic, we've dedicated an entire section to it—"Ain't Nothing Wrong with Using the Phone"—which you can find later in this chapter.)

Since one of the chief benefits of remote research methods is that it enables time-aware research, you'll ideally be able to observe users do what they were going to do before you called them, which makes establishing rapport and motivational inferences much easier. You'll find that your job is ultimately less about saying the right things and more about simply observing users as they go about their lives. Again, because watching people use interfaces in the real world for their own purposes is so different in terms of motivation, context, and actual behavior, time-aware remote research comes with a host of new moderating challenges, which is what this chapter is all about.

Introducing the Study

Let's start with the moment that users first pick up the phone. If participants are prerecruited, introducing yourself is simple and straightforward. If you're live recruiting, participants won't necessarily be expecting to be contacted, and you'll often have to do some quick orientation before you can begin the study. This first step is critical because not only are you about to ask users to do something pretty crazy—let a total stranger see their screen—but you're setting the tone for the whole phone conversation. You should introduce yourself, associate yourself with the survey that they've just filled out on the Web site (or remind them about the scheduled interview), and then ask them if they'd still like to participate in a 30–40 minute phone interview.

We've honed down this opening exchange so furiously over the past 10 years that it's practically the most efficient part of what we do. Feel free to modify it, of course, but we've found the exchange shown in Table 5.1 works really well.

TABLE 5.1 TYPICAL SESSION INTRODUCTORY FLOW

Moderator	Participant	Tips
Hi, can I speak to Bill Brown? Hi Bill; this is Nate, and I'm calling in regards to the: *[survey you filled out on ACME.com a few seconds ago]* *[interview we scheduled for today on ACME.com]*	"Wow, you guys are fast." *[Live recruits]* "Oh, that's right." *[Scheduled participants]*	About 90% of respondents answer the phone if you call right after they fill out the form. The longer you wait to call, the less likely they'll be to pick up.
How are you doing today?	"I'm doing fine, thanks."	
Is now a good time for the 30-minute phone interview?	"Umm, sure, yeah." *[70%]* "Sorry, now's not a good time." *[30%]*	A 70% success rate here, meaning that 30% of people just say no. If they don't have time or are otherwise unable to participate, thank them and end the session.
Great. You'll need to be able to talk on the phone and be on the Internet at the same time. Are you able to do that right now?	"Yes." *[80%]* "No." *[20%]*	An 80% success rate here. You'd be surprised how many people are at work and were planning on going to the break room or something.

TABLE 5.1 TYPICAL SESSION INTRODUCTORY FLOW
(CONTINUED)

Wonderful. We also ask that you install a small browser plug-in that lets us follow along with what you're doing onscreen just for the duration of this call. This will be completely removed at the end of the call and doesn't require any administrator privileges to install. Will that be okay?	"Sure, that's fine." *[70%]* "Hmm, I'm not sure. Can you explain more?" *[10%]* "Sounds complicated. Actually, I don't think I have time for this after all." *[10%]* "No. Go screw yourself." *[10%, so don't take it personally]*	Go along with whatever the users want here—no trying to persuade or convince people that they should use the screen sharing.
Now if you have a Web browser handy, I have a URL to read to you where you can give your written consent to participate today, and after you read that and fill it out, it will automatically take us to the page where we set up the screen sharing application. So, if you're ready, it's: *[Spell out the Web URL]*	"Okay, gotcha." *[80%]* "Can you spell that for me again?" *[18%]* "Where do I type in the address?" *[2%]*	Using the "T-as-in-Tiger, O-as-in-Orange" spelling trick is helpful here; Web addresses are always hard to hear over the phone. Some users will inevitably be computer inexperienced and have trouble setting up the screen sharing. Be patient; this is real user behavior, after all.

Some people express surprise at being called so soon after filling out an online survey, but this doesn't usually affect their willingness to participate one tiny bit. Just reassure them that you've been standing by to contact participants.

Assuming they agree, you'll need to explain about screen sharing, making sure they understand that (1) you'll be able to see everything they can see on their whole computer screen; (2) it's just for the duration of the session; and (3) you'll also be recording the screen sharing and conversation, just for research purposes.

Reluctant users may ask you about the testing process. Common questions include the following:

Q: Will you be able to take control of my computer?

A: No, Adobe Connect only allows us to see whatever you can see on your screen, and naturally we'll let you minimize or hide anything that's confidential.

Q: I'm a little busy now. Can you call me back later?

A1: [*If that's okay*]: Sure, when would you like to reschedule for? [Make note of time]

A2: [*If that's not okay*]: Tell you what—if you give us a time when you'd be available to talk, we'll see if that works with our testing schedule; if it does, we'll give you a call then. Does that sound okay? [*Make note of time*]

Q: I'm at work now, and I think my office firewall blocks downloads. Will it still work?

A: We can give it a try, if that's okay with you. That's usually not a problem, and it should take only a few minutes. What do you think?

Q: What's the purpose of this call again?

A: We're talking to visitors to the ACME Web site to get their feedback on a new version of the Web site. We'll have you show us how you would use it, tell us where you run into any difficulty, and we'll also be asking you a few questions about your experience. Does that sound okay?

Q: Something came up, and now I'm in a rush. Can we do this in just 20 minutes?

A: Unfortunately, we have to talk for the full 40 minutes to make sure we cover all our bases. Is there another time when you'd be able to talk to us?

NOTE **PRO TIP: DISMISSING RECRUITS**

It can be tough to tell recruits flat-out that the interview just isn't going to work out. Most researchers, out of politeness, naturally want to offer explanations, apologize, and talk more than is necessary. You don't want to do that. If you need to dismiss users (can't set up the screen sharing, don't have time to fully participate, aren't actually qualified to participate, etc.), and you've already spent up to 15 minutes trying to help them set up the study, and you have plenty of other qualified recruits that just came in via live recruiting, then dismiss the user. Use direct, certain, and friendly language to express that you can't go on with the session, citing the main reason without overexplaining:

"Unfortunately, the interview would have to begin right now..."

"I'm sorry to say that the screen sharing software is essential to the study..."

"Actually, we're currently looking to speak to people who are regular visitors to the ACME site..."

And then finish with something like this:

"...So we won't be able to conduct the interview. But thanks so much for responding, and I really appreciate your time. Have a good evening."

It's important to know that 1 in 1,000 survey respondents will try to insist on compensation for simply answering the phone or filling out the online screener. Since we like to treat users well, we'll sometimes offer a partial incentive or some token of appreciation, and the total cost to the study is usually negligible. But as long as you made it clear on the screener that incentives are given only to participants who *complete* the study, you're not obligated to do so.

Once you've walked participants through the installation process, you'll prepare them by introducing the purpose of the study and explaining broadly what participation entails. This step is important for keeping the participants from becoming confused, frustrated, or distracted later. At this point, they're usually assuming that you're going to tell them what to do and will be awaiting instructions from you.

First, you need to give a quick one-sentence description of what the study's about, and then—assuming you've designed a time-aware, live-recruited study, as we highly recommend—you'll tell them that you'd like them to do exactly what they were in the middle of doing when you called. (If some time has passed since they filled out the survey, you can ask them instead to do what they were in the middle of doing before they filled out the survey.) The very last thing you need to tell participants before starting is to "think aloud" as they're going about their tasks, so you can hear what's going through their minds. (For tips on how to word these statements exactly, see the facilitator guide walkthrough in Chapter 2.)

Unfortunately, sometimes you have to fight an uphill battle against user boredom and distraction. It's harder to keep people engaged over the telephone, especially when they're in their own homes or offices and (for all you know) may be watching TV, fiddling with a Rubik's Cube, etc. We try to give out instructions in small, measured chunks, rather than all at once—a strategy we call "progressive disclosure of information." If you catch yourself issuing instructions continuously for longer than 30 seconds, stop and engage users by asking if they have any questions, or have users perform some setup task (e.g., set up the screen sharing, open the interface) to maintain attention.

NOTE DON'T SPY

Users should always know before the session begins exactly the kind of access you'll have to their computer, whether you'll record the session, and what you'll do about the recording. If they have questions or reservations, don't try to pressure them into consenting. If you're live recruiting, you can always just snag another user.

Of course, it's tempting to look at all the juicy user info that's right there in front of you, but a few things are out of bounds for most purposes. For example, if users have carelessly left their email client open (despite your earlier instructions to hide any sensitive info), you shouldn't look though their personal email correspondences. Avoid pointing out any material that might make users feel as though they're being intrusively scrutinized or that has the potential for making an awkward situation. Examples of things you *don't* want to say: "Hey, I see there in your Web history that you just went shopping for boxer briefs!" "Want to show me what's in all those image files on your desktop?"

Privacy isn't just an ethical issue. If users even suspect that you're overstepping your bounds, they may feel anxious, untrusting, and defensive, and it'll shatter that all-important trust relationship that you must develop to get natural behavior and candid feedback.

Time-Aware Research and Going Off-Script

In the past, one of the central challenges of designing a traditional user research study was to come up with a valid task for users to complete: some activity that was typical of most users and realistic for users to complete without explicit instruction. Live recruiting, as you now know, makes this issue of "realism" moot because participants are screened based on the relevance of their natural tasks. Much of remote moderating, then, consists of listening to and observing users as they do their own thing.

With this method, you surrender a measure of strict uniformity in exchange for something that's potentially much more valuable, which is task validity. But this doesn't mean that participants can do literally anything they want for the entire session. As a moderator, you need to be able to distinguish which natural tasks are worthwhile and which ones will require you to step in and, well, moderate.

The best remote sessions are the ones in which you can address all the study goals without asking a single question. Ideally, in the process of accomplishing their own goals, users will use every part of the interface you're interested in and be able to narrate their interaction articulately the entire time. It's nice when this happens, but usually you'll have to do some coaxing to cover all your bases, and when that's necessary, we recommend a somewhat Socratic method: try to keep your questions and prompts as neutral as possible and focus on getting users to speak about whatever they're concentrating on. (A list of handy prompts can be found in the remote facilitator guide walkthrough in Chapter 2.)

NOTE DON'T INTERFERE WITH THE USER

Moderators are often terrified of inadequately covering everything that's been laid out in the moderator script. Since this is time-aware research, however, we encourage you to chill out. First of all, you *want* to see what users naturally do and don't do. Second, you can always go back later in the session to catch what you missed. The reason we so zealously endorse abandoning the facilitator guide and embracing improvisation is that we feel time-aware research is especially rife with opportunities to observe new and unexpected behaviors, much more often than you'd find in a lab setting. Some folks in the UX field have dubbed these unexpected behaviors as "emerging themes" or "emerging topics," and they come up all the time when you're speaking to users who are on their own computers and in their own physical environments. See "The Technological Ecosystem" in Chapter 7.

Although we encourage you to ditch the facilitator guide, it's still important to be ready for problems during the session. We've cranked out a list of handy things to say in the most common situations (see Table 5.2). They all come up sooner or later, and knowing what to say will help you project the appropriate air of serene moderator neutrality:

There are a number of situations in which you might not want to discourage users from the path that they're on, but you do want to speed them along. Most often it's when they read long blocks of text, pursue a goal entirely unrelated to the scope of the study, or (as is often the case with less tech-savvy users) can't figure out how to perform a basic task on the computer, such as enter an address into a browser bar. In the latter case, feel free to offer help; in the two former cases, you can try instead to speed up the process by asking them to skip over the rest of the task, if possible:

Thanks for showing me this. So, in the interest of your time, I was curious to see what you would do after you were finished reading this text over?

(Note how we say "*your* time"; this is just to emphasize that moving things along will keep the session from going over time, which is in *their* interest.)

TABLE 5.2 LIST OF HANDY TROUBLESHOOTING PHRASES

Situation	What You Say
Something goes wrong with your technical setup (screen sharing session disconnects, computer slows down, desk catches fire, etc.).	"Hey, can I put you on hold for just a minute? I've got to adjust something on my end." *[Place user on hold.]*
The phone connection is bad.	"You know, I think we might have a bad phone connection. Would you happen to be on a cell phone?" *[If so]* "Would you mind switching to a land line? Would that be possible?" *[If not]* "Let me call you back on a different line. I think my phone might be acting up. I'll just be a minute."
Your participant asks you a biasing question (e.g., "Can you tell me what this button here is supposed to do?").	"Hmm, that's a good question. First, can you tell me what your impression is?" OR, "You know, I'm actually not sure myself; I wasn't involved with the design. What would your best guess be?"
The user is taking an unduly long time on a relevant task.	"So, actually, in the interest of time, I'm curious to see what you'd do *after* you finished with this."
The user has become side-tracked on an irrelevant task.	"So, let me ask you to back up a little. Thanks for showing me that, by the way. Can you tell me a little more about what you were doing on that earlier screen?"

You may not cover all your testing objectives in a session, or you may want more info around a particular behavior. Sometimes, your user may have only a very specific task to perform at the Web site and may finish it early in the session. You should set aside a few minutes at the end of each session to assign predetermined tasks to fill in these gaps. When users tell you that they're finished, it's time to circle back to the tasks you've missed. This process is pretty much the same as standard usability research: prompt users to perform tasks, observe tasks, ask users about tasks—you know the drill. If you run out of predetermined tasks and you still have some time left over, ask your observers if they have any other questions or tasks they'd like to see users perform. Then you can wrap up things as usual.

Smart Note Taking

You could use plain old pen and paper to take basic notes during a remote session, if that's what works for you. But taking full advantage of remote research tools means taking notes on the computer. It can make analyzing the session videos and extracting the main themes of the research a lot easier. (Plus, it's cooler.) When taking notes, your main priorities are to do the following:

- Reduce the amount of distraction incurred by taking notes.

- Enhance the verbatim accuracy of the notes.

- Mark clearly when in the session the noted behaviors occurred. You want to avoid having to rewatch the sessions again in their entirety; doing so is time consuming and unnecessary if your notes are in shape.

And let's be clear—smart note taking isn't about pointlessly obsessing over time. An extra second spent timestamping or tagging your notes during live research will save you 1,000 minutes of analysis. (That's our official formula.) Reviewing hours of research video without timestamps is eye-straining and tedious, and nobody conducting research outside academia ever has time to review footage and log data after the fact.

If you've been in the UX game for a while, you probably already have some kind of shorthand method of manually timestamping. We like the plain old *[MM.SS]* format. If you're making video recordings of each of the sessions, it's important to time your notes relative to when the video began (instead of an absolute time, like "3:45 PM"). If necessary, you can download one of the many free lightweight desktop stopwatch programs online and start it as soon as your video recording starts. Whenever you make a note, copy the time on the stopwatch.

There are also more sophisticated alternatives to frenzied manual timestamping. Collaboration is one approach. Since remote research allows for a large number of observers who are also at their computers, you can prompt your observers in advance to make notes about user behaviors they're interested in. Using online chat rooms or collaboration tools like BaseCamp or Google Docs can provide a convenient outlet for observers to put notes and timestamps in the same place. Some research tools, such as UserVue's video marking function, allow observers to annotate the video recordings directly as the sessions are ongoing.

Are there any comprehensive solutions out there for collaborative, automatically timestamped, live note taking? NoteTaker for Mac has a built-in timestamp, but it's set to the system time, not relative to the session. There's nothing else as far as we know, but, happily, we've come up with a way to use instant messaging clients to do that. Our technique allows multiple people to work in the same document, with entries labeled by author and precise time noted automatically, down to the second.

The idea is to get everyone into the same IM chat room and then configure your IM program (Digsby, Adium, Trillian) so that it displays the exact timestamp every time anyone enters a note. Then set up a chat room and invite all the observers into it. (For step-by-step instructions on how to do all this, check our site at http://remoteusability.com.)

Once the session is over, everything that's been typed has been automatically logged by the IM client, so you have your timestamped transcript ready to go. You can either save or copy the chat room text to a document file—or better yet, an Excel spreadsheet, where you can easily annotate and organize each individual note (see Figures 5.1 and 5.2).

We know this technique is a little kludgy, but it's free, and it gets the job done. (We've also posted an Excel spreadsheet template with a custom-made macro script on our RemoteUsability Web site that does all the work of adjusting the timestamp automatically for you, but you may have to spend hours adjusting macro settings in Excel to allow a public file with an embedded macro to actually be allowed to run. Since that rampant outbreak of malicious macros in 1981, Microsoft has really clamped down on macro security in Excel files. Anyway.)

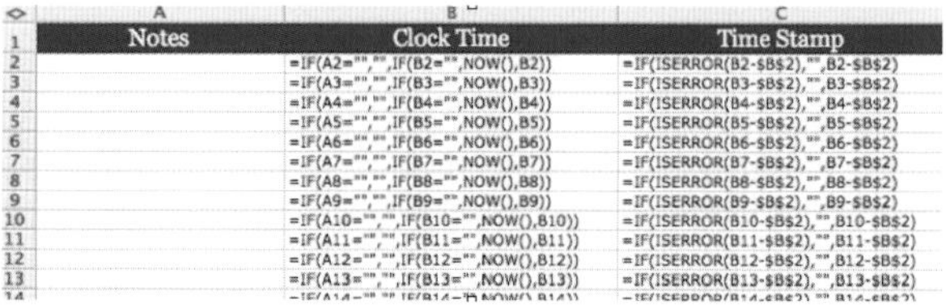

FIGURE 5.1

This is what our template automatic timestamping spreadsheet looks like before entering any info.

FIGURE 5.2
After entering your notes, the spreadsheet automatically enters the amount of time that has passed since the first note was entered. This way, you can track the times in your session recording when certain behaviors and quotes were made. (Remember to enter the first note right when the recording begins!)

	A	C
	Notes	Time Stamp
2	start	0:00:00
3	hmm, he uses google...	0:00:13
4	never sees home page	0:00:46
5	navigates straight to product page	0:00:54
6	misses central message	0:01:05

If the whole business of note taking just sounds tiring to you, you can always get someone else to do it for you. There are online services that will transcribe all the dialogue entirely. We personally like CastingWords (http://castingwords.com), although there are plenty of others. Transcription is a good option if you need all the dialogue transcribed verbatim within a limited time frame for some reason, or if you're just lazy and have lots of time and money you just don't need. The downsides to this approach are that it costs money (~$1.50/minute), you may have to extract the audio from the session video into an .mp3 or .wav file, and you'll have to wait for the transcription—the more you pay, the faster the service works.

Lastly, we're starting to see automatic transcription tools hit the scene. Adobe Premiere CS4, a video editing suite, has a feature that boasts automatic transcription of video files, with an interactive transcript feature by which you can jump to the part of the video in which a word was said simply by clicking the word. In theory, this obviates note taking; in practice, we've found that it's not very accurate at all, and the time it takes for the software to transcribe the video is roughly three times the length of the video. So you should probably stick to doing your own notes until the technology improves.

Testing Across Cultures

by Emilie Gould

*As we mentioned, one of the great benefits of remote testing is the ability to speak to users anywhere in the world where high-speed Internet is available. Here's **Emilie Gould**, expert in cross-cultural requirements, with a primer on the most important factors to keep in mind when moderating sessions with users abroad.*

Until the late 1990s, there was a tendency for interaction professionals to believe that everyone shared the same psychology and reacted to technology in the same general ways. However, research on problems in technology transfer has led to an expansion of the field of human factors and interaction design. More usability tests in more countries are revealing culturally based differences in the way people use computers.

Edward Hall developed the notion of "high- and low-context communication." High-context communication depends on the social context in which things are said rather than on the words that are used; low-context communication focuses more on the words. Suppose I ask you if you found this interface easy to use: you surmise that my feelings will be hurt if I tell you the truth, so you just say "Yes," but not much more. In this case, "yes" means "no," but I may not recognize that your concern for me should change my interpretation of the words and fail to recognize your lack of enthusiasm as part of your message.

Geert Hofstede identified five dimensions of culture:

- **Power Distance:** The extent to which people accept inequality in a culture. All societies tend to have hierarchies of some sort, but power distance does not refer to differences between people; rather, it refers to social tolerance for inequality.

- **Individualism and Collectivism:** Whether people act to maximize their own goals or sometimes restrict their actions in the interest of family, company, or in-group. However, individualists may choose to act in the interests of others, and collectivists have personal goals.

- **Masculinity and Femininity:** Whether people focus on actions and things or on emotions and relationships. This dimension covers motivations and social roles and, in Western society, is often associated with traditional gender roles.

Testing Across Cultures (continued)

- **Uncertainty Avoidance:** The amount of information that people need or want before they act. This is different than risk avoidance; people with high uncertainty avoidance will still take risks but want to know as much about the situation as they possibly can.

- **Long-Term vs. Short-Term Orientation:** The time horizon in which people act. Hofstede added this dimension to accommodate Chinese traditional values, which ground present actions in historical events or accept long delays to accomplish future goals.

When people apply a cultural frame to interaction design, they most often use Hofstede's dimensions to try to explain why people in another culture react in unexpected ways.

The way people use language is also subject to cultural differences. In many cultures in the Middle East, people use exaggeration to show their emotional commitment to what they are saying. Good things will be overpraised, and bad things will be strongly condemned; language is often embellished. Other cultures prefer understatement: in the United Kingdom and parts of Canada, simple words may be used in "nice" juxtapositions. Good is good and bad is bad; no more needs to be said.

Finally, culture affects nonverbal communication. The different elements of "paralanguage" (inflection, speed of speech, loudness, interruptibility) carry meaning. This dimension of intercultural difference relates back to our first topic: high- and low-context communication. It is not just who says something, but the way they use their voice to say it that must be interpreted to discern the correct meaning.

Don't let the fear of miscommunication stop you. Most cultures vary strongly on just one or two dimensions, and most people can overcome cultural differences by talking together first and establishing a mutually beneficial goal for their work together.

Emilie Gould's writing on cross-cultural moderating also appears in later sidebars in this chapter.

Shut up.
Listen.
Watch.

And take
good notes.

Working with Observers

With remote research, unlike in a lab, there's no need for two-way mirrors or omnidirectional cameras. Depending on the screen sharing tool you use, it's pretty easy to allow people to observe the sessions. On top of that, remote observers can easily use chat and IM to communicate with the moderator as the sessions are ongoing, giving their input on the recruiting process, alerting you to user behaviors you may have missed, and even helping you take notes during the session. If you're testing an interface whose technical complexity may be beyond your expertise (e.g., a glitchy prototype or a jargon-heavy interface), you can have an expert standing by during the session to troubleshoot and assist.

During the Session

There are a couple of things you should know to make cooperation smooth and useful during your remote session. First, make sure that the barrier between the observers and the study participants is airtight. There's nothing more disruptive to a study than having observers unmute their conference call line or have their chat made visible to the participant.

Observers should be briefed in advance about their participation in the research to make their contributions as useful and nondisruptive as possible (see "Preparing Observers for Testing" in Chapter 2), but you may find that observers occasionally overstep their bounds anyway. Their requests should be acknowledged ASAP, but you don't need to promise to act on every request. To avoid sounding rushed or irritated with observer requests, use concise and unambiguously friendly language to acknowledge them. Toss a smiley in there, if you're not ashamed: "Understood. I'll see what I can do! :)" Ask for clarification if the request is unclear: "Not quite sure I understand; can you clarify?" Ultimately, if there are too many requests to handle, the moderator should chat with the observers immediately following the session, mentioning any specific issues that kept them from fulfilling any requests: "Sorry I couldn't get to all your requests; it was important to leave enough time to cover our study goals. Feel free to keep sending them along, though!" The aim is to keep observers from overburdening the moderator, while not discouraging them from contributing usefully.

During the session, observers will naturally want to contribute, offer advice, and suggest questions they'd like to have asked (see Figure 5.3). That's fine, but the moderator should always serve as a filter between observers and participants. Always translate task requests into natural, conversational, and most importantly *neutral* language. For example, an observer may say, "Ask the user what he thinks of the navigation bar," but, of course, as a trained UX researcher, you know better than to ask for point-blank opinions. Instead, you could either simply wait for the user to interact with the navigation naturally, or you could ask questions geared toward eliciting a behavioral response: "What were you trying to do on this page?" "What would you do after this?" and so on. In any case, let the observers know that you're on top of the issue.

FIGURE 5.3
In this session, we projected the user's screen onto the wall and played the audio through speakers so that the observers could follow right along. Observers could pass notes or send IMs to the moderator as the session was ongoing.

You also may encounter the problem of observers who don't seem to be paying enough attention to the session or who remain silent. This won't necessarily harm the study, but it might indicate that the observers aren't aware of the active role they can play or are too timid to ask. If your observers are silent for a long time (~10 minutes), it's good practice to ask them if there's anything they're curious about or would like to have you probe further. If they say "no," carry on; if observers remain silent, you can follow up with a brief appraisal of the session, to encourage more observer discussion: "This user seems to have a lot of trouble logging in. I'll try to find out more about that."

Finally, avoid soliciting advice or over-relying on observers for guidance (e.g., "What should I ask the user now?"). Observer requests are a secondary tool for fulfilling the research objectives; the primary guidance, of course, should come from the moderator and from the research objectives, as embodied in the facilitator guide.

After the Session

If working with observers sounds stressful, including them is worthwhile for more reasons than just getting advice. Having stakeholders—developers, project managers, marketing people, whoever—observe and engage with the research can make them more invested in the outcome of the research. Findings are easier to understand and present, since observers can draw upon their own recollection of the sessions to place the findings in context. Stakeholders who are unfamiliar with remote methods (or UX research in general) and have doubts about their soundness and validity are often won over when they get to see how it happens. Since they gave their input throughout the process, they're less apt to question the researchers' approach.

Short debriefing sessions after each session are useful, letting observers fully elaborate on things they noted during the study. When testing is finished, observers can join in on brainstorming sessions and contribute any insights they may have gathered as outside observers. This is also a good opportunity to catch any mistaken assumptions your observers may have about the research. They might have interpreted a user's offhand comment that "This login process is easy" to mean that there are no problems with the login process, whereas you spotted dozens of behaviors to the contrary. You can then be prepared to address these assumptions as you're drafting your report.

When you finally present the findings, it's nice to acknowledge the observers' contributions to the project, specifically calling out instances when they helped. These acknowledgments make it clear to observers how they benefited the project, increasing their investment and confidence in the research. That makes it more likely the findings will be applied or possibly even convince observers that a stronger organizational focus on research is required, creating more opportunities for good research.

Testing Across Cultures: Interpreters and Cooperation

by Emilie Gould

Negotiating language use can be a major problem when working with participants who speak a different language from your own. Including an interpreter in the conversation may help you get your meaning across, but it brings several complications. First, your interpreter may not understand how important it is to avoid leading the participant; in trying to make your meaning clear, the interpreter may be too clear. Second, the participant now has two people to deal with: you and the interpreter. Instead of being able to project yourself as a peer, you now have another doing your bidding. This will make you seem more important and powerful than if you were speaking yourself. This may make the participant more likely to self-censor comments that might offend you.

Unlikely? Unfortunately not. Interaction researchers in both China and Malaysia have reported that participants often choose to preserve harmony when commenting on problematic software. They may have a horrible experience but say that the software was fine—to save your face. This is a good example of power distance and collectivism at work. Your only option is to establish that you are not the developer of the software, and you have nothing personally at stake, but that can be a hard sell. You may need to persuade participants that you need their help—putting yourself in a dependent position—and will not be offended by their comments to get closer to their real feelings.

Or you can just watch and see what happens when the participant uses the interface. Instead of depending on language, focus on actions and missteps for most of your information. [We prefer this approach. --Ed.]

Quiet, Chatty, Bored, Drunk, and Mean

You have a problem participant. Dang. Usually, you can catch these people right away with proper recruiting, but every now and again you'll be speaking to someone who seems like a decent enough participant at first and then becomes tough to communicate with once the session has begun. Your treatment of these participants will vary based on the study goals and also the nature and severity of the issue.

Quiet participants are the most common type of problem participant, but thankfully they're also the easiest to handle. Nine times out of ten, the issue will be that users get too engaged in their task and forget to think aloud in a way that helps you follow along. This is actually a good thing because it shows that they're engaged in a task they care about. You want to strike a balance between engagement and talking so that users are speaking undeliberatively about what they're feeling and doing, and what problems they're facing in the moment, rather than their *opinions* about the interface. In cases like this, all you have to do is encourage users to speak up. Here are a few polite prompts you can use:

- "So, tell me what you're trying to do here."

- "What are you trying to get done right now?"

- "How does this *[part/page]* compare with what you were expecting?"

- *[If the user falls quiet repeatedly]*: "And by the way, if you could just let me know what's going through your head as you use the site..."

Some users will take more encouragement than others, but that's life in the streets.

Other times, you're dealing with a bored participant who doesn't seem engaged *and* isn't speaking. Symptoms include minimal time spent on each task, flat conversational tone, and consistently short responses to your questions ("It's fine," "It's okay, I guess." "I dunno"). Here's where you'll use your moderator skills to come up with questions that are both very specific and hard to give short responses to. Ask about specific onscreen behaviors:

- "I noticed that you just hesitated a bit before clicking on that button. Can you tell me why?"

- "Why don't we back up a bit? I was curious about what drew your attention to the tab you just clicked on?"

- "Before we move on from here, I wanted to ask you about this part a bit more. What do you think about the range of choices they give you here? Is anything missing?"

On top of that, you can try to examine the *reasons* participants may be bored. Boredom often indicates that they're not doing a task they care about—i.e., a natural task. Since the point of time-aware research is precisely to study people who are doing something they're intrinsically motivated to do, you may have recruited someone whose task was slightly out of sync with the goals of the study, and that may mean you need to screen the participants more carefully. If you find yourself talking to several bored users, however, you may have a larger problem on your hands— namely, that *nobody* cares about doing the task you're studying. Don't panic; having this type of response just means you might have to rethink your facilitator guide, and even your research goals, on the fly. Try to keep the focus of the study on watching users accomplish objectives that matter to them; you'll get behavior that's both more natural and more engaged.

Besides bored or quiet users, you can also have users who are very chatty and try to tell you their life stories. You can hear them leaning back in their chairs to have a nice chat on the telephone. When this begins to happen, gently interrupt them when they reach the end of a sentence and try to refocus them on a task: "Can I interrupt you? Sorry, I was actually curious if you could...." If it becomes a repeat problem, mention how much time is remaining in the session: "We have about 10 minutes to go, and so to keep from running over our time, I just wanted to make sure we got through the whole process here...." Be polite but persistent; don't let users interrupt themselves with chatting.

Testing Across Cultures: Quiet Users

by Emilie Gould

Quiet participants may be quiet because they feel uncomfortable speaking while acting. That notion appears unlikely to Americans trained to speak up in school and "give it your best shot," but people in many, many countries hold their tongue until they can be sure they have it right and will not say the wrong thing. Is one system better than the other? You've probably run into participants who seem to spout off illogically. By contrast, many Asians and northern Europeans will take some time to contemplate their options before acting. Then they can tell you about the interface. But not before.

If you want to ask these participants why they did what they did, you should consider taking a few minutes at the end of the test for retrospective comments. Review their actions and ask them to say what they were feeling after the fact.

This strategy has an added benefit if you are working with people who are not completely fluent in the language of the study. Rather than distract their attention from the task by asking them to try to "think aloud," let them do the task and talk about their actions later. Cognitive researchers know that people can do only one thing at a time well. Generally, when we think we are multitasking, we are really just rapidly shifting focus. The harder the two tasks—like performing a task under scrutiny and searching for technical words in a second language—the more likely people will make errors that would not occur in real life.

Now here's something interesting. Since you may be calling people in different time zones—occasionally after Miller Time—you may reach some participants who are perfectly qualified, are willing to participate, and happen to have, you know, *had a few*. Which is not as bad as it sounds: people really *do* use computers in all kinds of chemical states, and we only think this is a problem if (1) it gets in the way of communicating with the users, or (2) it affects their ability to actually complete the task. Otherwise, if this is really the way they'd use their computer, we see no reason to exclude them. This is time-aware research at its best, folks.

Inevitably, you'll encounter users who have an attitude, want to cheat the system, have an axe to grind, or are hostile. We've seen every variation of this under the sun. One guy tried to impersonate his son to participate in the study twice; a handful of users have made inelegant passes at our moderators; and then there are profanities, sarcastic put-downs, and deliberate heel-dragging.

Our take on this? The world is what it is; users who are abusive to the moderator, or refuse to follow the basic terms of the study, have no place in your study. We authorize our moderators to terminate sessions immediately if they feel legitimately harassed. If we feel that we were able to get some useful insight from a participant who later becomes abusive, we might feel inclined to offer a partial incentive to the users, proportionate to the length of the terminated session. If we're not feeling so charitable, we just say:

> "I'm sorry, but I don't think we'll be able to complete the study without your cooperation. I'm afraid that we won't be able to offer you the Amazon gift certificate, since, as we mentioned in our recruiting survey, only users who complete the study receive the incentive. Have a good day."

And good riddance. This approach, of course, should be used only in the direst of circumstances, when it's unambiguously clear that the participant is a stinker. When the time comes, however, don't hesitate to drop the hammer.

Ain't Nothing Wrong with Using the Phone

For some UX veterans (see Andy Budd's interview in Chapter 1), the notion of talking to users over the phone is appalling. How are you supposed to build empathy with users if you don't have body language? How can you see what users are thinking if you can't see their facial expressions? Many practitioners balk at doing remote research for this reason alone. But we're here to tell you once and for all that there's nothing wrong with speaking over the phone for most user research. By now most people are perfectly comfortable with expressing themselves over the phone and can adapt their vocal tone to make their emotional cues and meanings understood.

In terms of communicating with the participant, the major difference between in-person and remote testing is that you won't be able to rely as much on body language as you might normally, for purposes of establishing trust and cutting through conversational tension. This makes the moderator's responsibilities more complex, since your job isn't to express yourself naturally, but to encourage users to speak their minds; translating the lab demeanor to the phone just takes some awareness of how to use your voice to moderate.

Take, for example, the habit of "active listening"—the practice of regularly nodding and saying "mm-hmm" to demonstrate that you get what users are saying. In person, nodding your head and repeating "gotcha" may encourage users to keep going, but over the phone, the moderator should cut back on the active listening "uh-huh"s and "gotcha"s because it actually encourages users to wrap up what they're saying. Don't be afraid to just sit back and listen—even allow a few awkward silences—to let users gather their thoughts and say everything they have to say.

Related to active listening is "reflecting," or paraphrasing or repeating things that users have just said to clarify or be sure that you've understood their meaning. This technique is somewhat risky for two reasons: (1) it almost always has the same suppressive effect as active listening; and (2) if your paraphrase is inaccurate, it can lead users to agreeing to propositions or coming up with ideas that they may not have otherwise. We've found that a better alternative is to begin as if you're going to paraphrase them but then have them do the bulk of the work by trailing off and letting them fill in their own thoughts, for example.:

> *Participant*: ...and so I decided to click on that link to go to the next page.

> *Moderator*: Okay, so let me get this straight; first, you saw the link, and... what, again?

> *Participant:* I saw the link, and I thought to myself... *[paraphrases self]*

Since people often paraphrase when repeating what they've just said, this strategy achieves the same end as reflecting, with less moderator bias.

Remember that over the phone, all you are is your voice, which means that any emotions you express vocally will have a heavier influence. The big thing to be careful about is staying emotionally on point. If you're having light banter with participants early in the session, feel free to talk and laugh along with the users; that helps cut down any self-consciousness that may affect their natural behavior. When dealing with users in a study that may be emotionally fraught for whatever reason (as we recently had to, for a study of cancer patients), don't shy away from expressing polite, sincere sympathy. However, when users get around to discussing anything related to the goals of the study, *don't* try to match their enthusiasm or frustration, but instead maintain a friendly, neutral tone. It makes an audible difference if you're actually feeling comfortable and you're using your own body language; it's easier to smile than to try to sound as though you're smiling.

Ultimately, no matter how much we try to tell you that talking to users over the phone is just as effective and engaged as in-person research, nothing's going to make the case better than trying it for yourself. You can easily just shrug and wring your hands about how this type of research is not going to work, but don't hate it until you've given it a shot.

> **NOTE** ARE YOU GOOD AT MULTITASKING?
>
> You will do a lot of multitasking. Aside from the usual moderator tasks (taking notes, asking questions, referring to the facilitator guide), you'll also be dividing your attention between IM chat conversations with your observers, which is a substantial attention-sink (see Figure 5.4). If you're not confident in your ability to multitask, consider easing back on the note taking and simply refer to the session recordings later. Working this way is time consuming but better than getting distracted and spoiling a session.

FIGURE 5.4
The moderator's desktop: IM, notes, screen sharing, script, and Web browser (and sometimes recording). A second monitor helps keep things much less cluttered.

Wrapping Up

When all's said and done, most users care about only two things: how they'll get paid and how to get the screen sharing stuff off their computer. Wrapping up a study should cover at least these two points, which usually takes no longer than three minutes with the script we gave you in Chapter 2. Very rarely are there any snags in this step; if users have additional questions that you either can't or don't have time to answer right away, feel free to tell them that you'll email them with an answer as soon as possible. And then you're done. Exhale.

Chapter Summary

- When introducing the study, quickly establish who you are and what the study is about. Then introduce the terms and instructions of the study in small chunks, so the participant doesn't zone out.

- Feel free to go off-script if users do unexpected things relevant to understanding how they use your product, but know how to keep them on track and stay within the allotted time.

- You can take timestamped notes more efficiently with jury-rigged spreadsheets and programs; transcription services and voice-to-text software are also options.

- Set your observers' expectations for the session ahead of time and work with them to get their feedback and insight as sessions are ongoing.

- You'll inevitably encounter problem users. Use good moderating skills to deal with quiet, shy, and bored users, decide whether chemically altered users will be viable participants, and politely dismiss rude or uncooperative users.

- Don't worry about not seeing the users' faces; a lot of expression comes through in vocal tone, and it is behavior you're most interested in. Adapt your speaking manner to the telephone in order to encourage natural user behavior.

CHAPTER 6

Automated Research

W<span></span>e now turn to a different branch of remote user research: *automated research*, also known as *unmoderated* or *asynchronous* research. Unlike moderated research, automated research does not involve any direct interaction or conversation between the researcher and the participants; instead, you use online tools and services to collect information from users automatically from the Web. You can conduct automated research to get feedback from a large number of participants about behavior on a specific set of tasks. The majority of automated research is performed on live Web sites, since most automated research tools track Web interaction, but other tools have users interacting with static images or performing conceptual tasks, like card sorts and games, which can give you different kinds of feedback.

If that all sounds like jargon, think about automated research as "task-specific Web analytics." Automated research is similar to Web analytics in that it tracks the behavior of large, statistically significant samples, but it differs in that you give participants tasks or exercises to complete to look at the data through a less general lens than standard Web analytics. With Google Analytics, for instance, you can see how long people spent on different pages of your site or the average time spent on a page, but you have no idea why. With automated remote research, you use a tool that puts the users' behavior in the context of a certain task, such as "creating an account." The kinds of log data you get are similar, but since they're specific to a task, the findings are more targeted.

In this chapter, we'll talk about how automated studies are structured, what different kinds of automated research there are, how to recruit a large number of users to participate, and how to design the tasks.

The Structure of an Automated Study

The centerpiece of moderated research was the testing session: everything you did was to ensure that things went smoothly while you observed and spoke with the users. In an automated study, you don't interact with users at all, so the study's entire emphasis is on the way you set up everything.

Generally, the structure of an automated study looks something like this:

1. Choose an automated research method based on what you want to find out.

2. Select the tool or service that best supports your research goals.

3. Design a set of tasks or prompts for the user to complete with that tool.

4. Recruit a pool of potential participants.

5. Send email invitations to valid recruits.

6. Repeat steps 4 and 5 until user quotas are met.

7. Analyze the results and pay out the incentives.

Automated Research Doesn't Mean You Get to Be Lazy

There's been a line of thinking around automated research in that people conceive of automated research as some futuristic time-saving convenience from *The Jetsons*, and say, "Oh, cool, so it's like moderated research, except it's faster, I get way more users, and I don't have to do anything? Sign me up! Here's my expense account card!"

On the contrary, automated research is just plain different from moderated research, and by no means a replacement. While it's great to get a wide view on how a large number of people use your site, the main shortcoming of automated methods is that, compared with moderated methods, they offer less behavioral context. Although automated tools usually offer a few ways to gain insight on what users are thinking as they interact (prompts about why they performed or abandoned tasks, mouse movement recording, and so on), you lose the ability to ask questions about unexpected behaviors, observe how users are using other Web sites and artifacts in conjunction with your site, and hear users' spontaneous think-aloud comments, which usually differ widely from their responses to specific questions.

Furthermore, automated methods are limited by the fact that the tasks are prompted. When you ask users to do something, it's like asking them to solve a puzzle; they take all the time they need to reflect on the "right answer" and ask themselves, "What would I do

> ## Automated Research Doesn't Mean You Get to Be Lazy (continued)
>
> here?" On the other hand, people who just come to the Web site on their own initiative to accomplish their own goals will exhibit natural, self-directed behavior, without thinking about their performance. (And Web analytics, which *does* track natural user behavior, lacks the ability to control the user group to a fine-grained extent.) The qualitative stuff is what gives you the deepest insight into *why* users are doing what they do on the Web site, and for now, the best way to get that information is by having a trained moderator observe and react to user behavior.
>
> And on top of all that, automated research is best when you run multiple studies over time. The full usefulness shines through when you conduct studies at regular intervals, since they'll provide the benchmarks by which you can track how well your Web site is going along. So you may even need to do several.
>
> For all these reasons, we caution you against thinking of automated research as any kind of a shortcut or time-saving or money-saving technique or even a way to get around doing moderated research, because it isn't. Instead, automated research is a way to answer certain targeted questions about your interface. It's great for gathering concrete data about which of a handful of designs performs better or how well your interface supports a specific task or where users are clicking most. We see automated research as a complement to, not a replacement for, moderated research. Automated research is cool, but it isn't *everything*.

Recruiting for Automated Studies

For an automated study, you'll recruit many more people than you would for a moderated study. We've done automated card sorts with as few as 25 users, but a lot of researchers are drawn to automated research because of how easy it is to recruit a large, statistically significant sample, and for those purposes, you'll probably want something closer to 100 or more.

The Web recruiting methods we discussed in Chapter 3 still largely apply. On top of the screening questions, the recruiting form or screener you use will need to ask for the respondents' consent to be contacted by email, as well as the respondents' email addresses; real names and phone numbers won't be necessary. As responses come in, you'll send out emails to all qualified users; some automated research tools have built-in email recruiting to help you manage who's responded to the study.

Sample Automated Recruiting Email

Dear ACME.com user:

We are working to improve the ACME.com homepage; based on the survey you filled out recently at acme.com, we would like to invite you to participate in an online usability study. It takes about 15–20 minutes to complete.

-- Click on the link at the bottom of this email to begin the study. --

Every 10th person to complete the study will receive a $75 Amazon gift certificate for participating. This invitation is for you only; it can't be forwarded to another person, and the study can't be completed more than once by an individual. If the survey is closed, we have already received our maximum number of responses.

-- http://shorturl.com/go --

If you have any questions, please email researcher@acme.com. Thank you for your participation!

Not everyone who receives the recruiting email will complete the study, so you'll probably have to send out several rounds of invitations. You won't have to be standing by to contact users live, as you ideally would in a moderated study, but the greater number of users means you'll have to wait longer, and if your Web site doesn't have a ton of traffic, recruiting the way you would for a moderated study may not be feasible. Using this approach will help to keep your recruiting criteria broad and your recruiting screener short (remember, you won't have to ask for a phone number or name), but most importantly, you need to come up with an incentive scheme that will bring in a lot of users without breaking the bank.

Offering $75 to each of your 100+ participants isn't necessary, mercifully. Two options are to either offer a very small incentive (~$5) to all participants or to offer a 1-in-10 chance of receiving a standard $75 incentive. (There's no reason it has to be 1-in-10; that's just what's worked for us.) We usually go with the latter simply because it's less of a pain to manage and deliver 10 payments than 100. However, guaranteeing a small amount of money like $5 is often more effective than saying "Every 10th respondent earns $75." In online research, anything that seems like a shady contest or promotion sends people running for the hills. As always, you can offer whatever merchandise or in-house discount incentive you have on hand, if you think people will be interested. You'll need to keep a close eye on the number of invites you send out so that you can stay true to your word about the reward rate. You also need to spell out that the recruiting screener is *not* the automated study itself.

Are there any alternatives if do-it-yourself Web recruiting isn't an option? Again, we advise against resorting to the ever-tempting bastion of eager fake participants that is craigslist, but sometimes that's the only way. Recruiting from a list of your email contacts and mailing lists is still possible, but since you're sending out so many more invites, you'll risk running through your incentives even more quickly than you would for a moderated study. Also, you probably won't be able to get "first-time" impressions this way, unless you're testing a prototype. And then there are market research recruiting companies, and some automated research companies like UserZoom and Keynote offer panel recruiting services with their tool, which are pools of general-purpose standby users who get paid to do automated studies—in other words, professional research participants who get paid to fabricate opinions all the time. As with any indiscriminate recruiting method, these participants may not care about your product or service at all, and in our experience, they tend to care more about getting paid than they do about giving reasoned, deliberate feedback. We wouldn't recommend using such services. If you can recruit from your Web site, you should.

Sorry, Nerds

Where automated research is concerned, in this book we're going to focus on helping you decide what kind of automated tool you want to use, guiding you through setup and recruiting, and (in Chapter 7) helping you quickly interpret the findings you get. Consequently, there are a bunch of things we've left out because we felt that they were unnecessarily complex for addressing simple questions. We'll help you choose and design the right kinds of tasks for your users to perform, but we don't get into number-crunchy regression/conjoint/factor/quadrant analysis, optimization, or any of the multivariate stuff you learned in AP Calc. While rigorous analysis of findings is fine, we didn't feel that this was the right place to hit you in the face with a ton of math, and besides, we want nonmath people to be able to do simple automated research, too.

But maybe you are into that kind of thing! There are two books you should check out. One is called *Measuring the User Experience* by Tom Tullis and Bill Albert. In this book you'll learn all about the kind of data analysis we sheepishly avoid. The book is available from Morgan Kaufmann Publishers. The other book, which goes into fine-grained, advanced automated research techniques, is called *Beyond the Usability Lab: Conducting Large-Scale User Experience Studies*, by Bill Albert, Tom Tullis, and Donna Tedesco, who is the technical editor for the automated research chapters in this book. The latter book, to be published in January 2010 by Morgan Kaufmann Publishers, even contains a case study from Yours Truly. Learn more about both of these books at www.measuringuserexperience.com.

Different Kinds of Automated Research

Automated research encompasses many different methods, which are fragmented across a mess of different tools. It's common to have different methods in a single study; you might have users answer a few survey questions and then perform some tasks. Each method has different dimensions (see Figure 6.1). You can gather feedback that's opinion based (i.e., market research) or behavioral, open-ended or closed-ended, concrete or abstract, qualitative or quantitative.

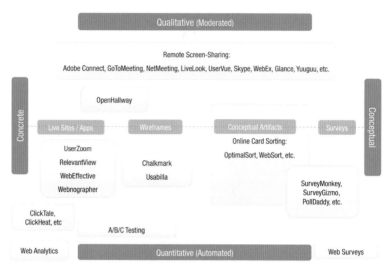

FIGURE 6.1

Various dimensions of automated research. The Concrete/Conceptual axis indicates how close user tasks represent actual behavior on a completed interface, while the Qualitative/Quantitative axis is more or less synonymous with whether or not the research is moderated or automated.

Now we'll discuss the three different methods you're most likely to encounter: task elicitation, card sorting, and surveys. Even though we address the three methods separately, we'll use case studies to illustrate the ways you can adjust, combine, and tailor different methods to suit your needs. While these three alone don't cover all automated research, they constitute a large portion of the automated research that's currently done, and they are nicely distributed among the various axes described previously. We'll also throw in a list of other various methods that are out there.

Task Elicitation

Task elicitation refers to automated tools that prompt users to perform a task and then record the users' behavior in response. It's the standard method of collecting behavioral data about a specific task on a Web site. An example would be a tool that asks, "Suppose you were looking to buy an ACME Googlydooter. Where on this page would you click to find information about the Googlydooter?" It then records where on the Web site the user clicks.

There are many variations on this type of tool: some prompt you to click on a part of the screen and then record the click; some prompt you to perform a task and then record your behavior as you navigate through and interact with the Web site; some ask you to manually indicate when you've accomplished or abandoned a task; and so on. If you want to learn how users interact with some part of your Web site, task elicitation is the way to go. It's a flexible way of getting a specific answer to common usability issues, such as the following:

- Can users perform a certain task on the Web site?
- Do they have difficulty performing the task?
- How many people abandon the task, and where?
- What's the most common path users take through the interface to complete a task?
- Where do users go to perform the task, and where do they *expect* to go?
- Do users know where to find the relevant links/info in the interface to be able to complete a certain task?

Fundamentally, designing a task elicitation study is about asking the users to do something on your site, simple as that. There are different ways to design the prompts, however, and plenty of common mistakes to avoid.

Don't Lead the Users

When writing a task, you want to be absolutely clear about what the users should do. If you want your study to have a shred of validity, you must *not* lead the users with the task prompt. It's really easy to do this by accident; we know, because we've messed it up before. In a study we conducted for UCSF Medical Center using the automated service UserZoom (see Figure 6.2), we gave users the following prompt:

> "For this task, please assume that your child has been successfully treated for a brain aneurysm. (A brain aneurysm is a balloon- or bubble-like growth usually located on an artery at the base of the brain.) It is time to make a follow-up appointment. Your task is to **find the phone number for returning pediatric cerebrovascular patients to call for an appointment.**"

FIGURE 6.2
UserZoom screenshot from our UCSF study. User instructions are persistently displayed in a browser bar at the bottom of the screen.

Our big mistake here was in supplying the users with the exact term "pediatric cerebrovascular." Since UserZoom employs a "browser bar" that frames the Web page while users go about this task—and since "pediatric cerebrovascular" isn't easy to spell—it's likely that a portion of the users in the study copied the term from the browser bar directly into the Web site's search bar or used their browser's Find feature to locate those two exact words, making the task artificially easy. Testing the accuracy of the search engine was part of the study goals, and so giving users the exact spelling ruined any chance of getting data on how the search engine was failing to return results for common misspellings. Furthermore, our prompt also implicitly informed users that there is a special line for pediatric cerebrovascular patients to call, which would likely prevent many of them from attempting different ways to make an appointment.

Find the Right Level of Task Specificity

It's easy to assume that the more precise you are, the more valid the task will be, but actually you need to design tasks using prompts that will suit the users' way of thinking about the task. The more general the phrasing of the question, the broader the nature of the answer will be. Let's tweak that erroneous last sentence in the previous example, to illustrate three different levels of specificity that accomplish three different goals. Starting with the most general:

> "Your task is to find the information you'd use to schedule a follow-up appointment."

By keeping the task as open-ended as possible, you could see whether users attempt different methods of scheduling an appointment, including email or contact form. By using a follow-up multiple-choice question like "What kind of information did you find?" you can then determine what percentages of users even look for a phone number to begin with (a growing concern in the Internet age). Here's another possibility:

> "Your task is to find the phone number to call to schedule a follow-up appointment."

With this phrasing, you'd be able to get at questions like this: Do users know there are patient-specific numbers? What percentage of all visitors (including the ones who aren't aware of it) are able to find the special number? Here's a third variation:

"Your task is to find the phone number provided for parents of children who have undergone brain surgery to schedule a follow-up appointment."

This question communicates the task without giving users any special terms that would make the task easier. You'd be able to determine how easy the task is for people who know there is a special number to call but don't necessarily know where on the Web site to find it.

As you can see, the information you disclose when you ask the users to do something has a huge impact on the sort of information you'll get from the findings, so pay attention.

Use Both Qualitative and Quantitative Methods

Since, as we've mentioned, automated methods suffer from a dearth of rich context—you know where people click but can't be sure why—you should try to get at your goals from many different qualitative and quantitative angles. You can ask open-ended questions addressing how the users felt about the task immediately after performing it, or you can ask closed-ended questions having the users rate the difficulty of the task. You can also verify that the users' completed the task successfully with a "quiz" question. If you take that approach, you'll need to inform the users in advance that they will be required to write down the info they find to be able to answer the question later. (Some tools like RelevantView allow the users to go back into the interface to look for the answer, and the time spent doing this is noted.)

We took many of these approaches in our UCSF study. Users were prompted to indicate that they were finished with their task by clicking either a "Success" or "Abandon" button. After each task, the users were asked a single multiple-choice question to verify the task success. (For the previous question, it was "Please select the phone number you would call as a returning patient to contact the Center.") Next, we provided a Likert scale (rate from 1 to 7) for users to reply to the following question:

Please indicate how easy or difficult it was for you to complete this task.

Then we provided more Likert scales to indicate agreement or disagreement with the following statements:

- It was clear how to start searching.

- Finding the correct phone number was easy.

- I felt satisfied with my results at the end of this task.

- Doing this task increased my confidence in the UCSF Medical Center.

- This task took a reasonable amount of time to complete.

After those questions, we asked an open-ended follow-up question: "If you have answered 1, 2, or 3 in any of the statements, please explain why." The purpose of this question was to get a close sense of the users' perspectives on any negative experiences they might have had. (Some tools have branching-logic capabilities, so you would be able to present the follow-up question automatically to whoever answers 1, 2, or 3.)

Finally, we offered a list of "Check all that apply" statements:

- I didn't experience any problems.

- Finding the information took too much time.

- The information presented was not clear.

- The information presented was incomplete.

- It was difficult to know what steps or path to take to take to find the correct information.

- The information, options, or menus were difficult to understand.

- I didn't find the information in the place I was expecting to find it.

- The links were unclear.

- Links didn't take me to the information/section I was expecting.

- I couldn't figure out what to search for.

- I had trouble spelling or typing something I wanted to search for.

- Search results didn't meet my expectations.

- Other (open-ended response).

In this way, we used many different means to get at the same issue, instead of being completely in the dark about why users clicked where they clicked.

NOTE SAMPLE OPEN-ENDED RESPONSES

Here are some examples of comments we received in response to the open-ended questions to the pediatric cerebrovascular appointment question:

- "I could not find the specialty in the pull-down menu on the appt. section."

- "Too much text to read—I'm interested in making an appointment, not reading a book."

- "Since I couldn't set an appt. by specialty, I thought that I might look for the facility—the opportunity was not there."

- "Of the numbers provided, all at the bottom of the page, none were specific to a children's hospital."

- "In the end, I'd rather call the main desk than fish around this page—no doubt both will be equally frustrating."

- "I didn't finish the task because I felt like I had exhausted all my options."

- "Could not find the information under 'appointments.' Tried searching, even though I would not do that in real life. I would give up immediately and call the main hospital number, or any number for that matter, to get someone on the phone and ask."

These comments give you an idea of how users feel about the task with some specificity, giving you some context to their behavior. (Not all comments are helpful, however, and as always, you should be wary of self-reported feedback.)

Case Study: Club Med and Beaches.com Competitive Analysis

by Carol Farnsworth

Carol Farnsworth, Senior Director of Customer Experience for Keynote Systems (makers of the automated usability tool WebEffective), has been conducting user research for over 10 years, recently focusing on Web usability research. Carol taught usability methods and testing courses at Stanford University, serving as a faculty member in the Information/Web Technology department. She holds professional affiliations with the Usability Professionals Association, BayCHI, and AeA.

On behalf of Club Med, I conducted a task-based automated competitive study comparing the Club Med and Beaches sites, using Keynote's WebEffective tool. I chose this methodology because I wanted to study user behaviors in their natural setting. I also wanted a large sample size, unlimited geographic reach, the ability to segment the data, and limited group influence or moderator bias. Because of the expense of development, few clients want to implement a new Web site design until statistically significant numbers of their customers and prospects have interacted with prototype designs.

The goals of the study were to understand the relative strengths and weaknesses of the Club Med and Beaches sites, and to use the findings to improve the Club Med site and increase the conversion rate.

SETUP

The Club Med vs. Beaches study was a between-groups study where panelists completed four tasks on one of the two sites (see Figure 6.3). One hundred panelists were recruited for the study from a third-party market research firm, each meeting the following criteria:

- Had a prior vacation at an all-inclusive resort

- Have children under 18

- Equal gender mix

AUTOMATED RESEARCH   135

Case Study: Club Med and Beaches.com Competitive Analysis (continued)

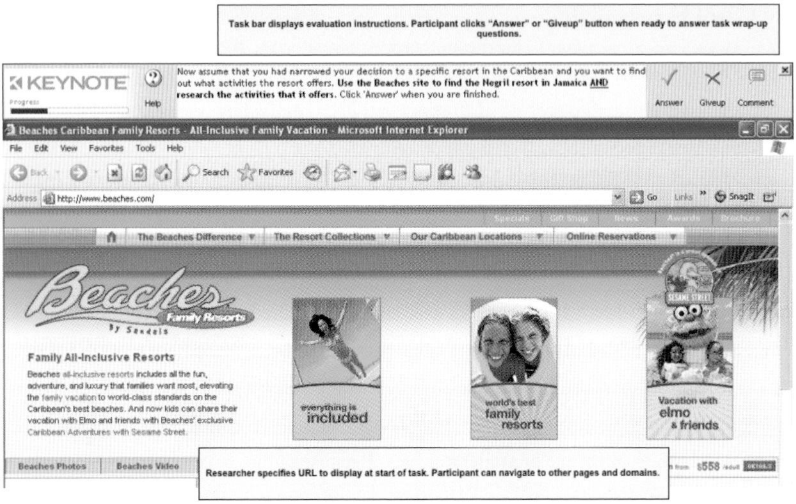

FIGURE 6.3
An example task given to participants: finding and researching the Negril resort in Jamaica.

STUDY TASKS

Introduction and Background Questions

Task 1: *Homepage evaluation*

Task 2: *Find a resort (open browsing)*

Task 3: *Find a specific resort (directed browsing)*

Task 4: *Booking process*

Wrap-up

Here's an example of one of the tasks and follow-up questions for Task 3: *Find a specific resort (directed browsing)*:

- Now assume that you had narrowed your decision to a specific resort in the Caribbean and you want to find out what activities the resort offers. Use the Club

Case Study: Club Med and Beaches.com Competitive Analysis (continued)

Med site to find the Punta Cana resort in the Dominican Republic AND research the activities that it offers.

- Click "Answer" when you are finished.

We asked the participants if they were successful to obtain self-reported success:

- Were you able to find the activities available at the Punta Cana resort? [Yes/No/I'm not sure]

Also, if the participant reached one of two URLs, WebEffective indicated a task success, displaying a message:

- Congratulations! You have successfully completed this task. Please click "continue" to proceed.

We subsequently asked several closed- and open-ended questions, including

- While trying to complete this task, how satisfied were you with your experience on the Beaches site? [Rate 1 to 7; 1 = Not at all satisfied, 7 = Extremely satisfied]

- What aspects of the Beaches site, if any, did you NOT like while looking for activities for a specific resort? [Text response]

- Which of the following problems or frustrations, if any, did you encounter while trying to find activities for a specific resort? [Check all that apply: Difficult to click around between activities./Difficult to navigate the site./Site was slow./Layout was confusing./Site errors, pages not loading./Too much clutter on the pages./Too many steps required to find a specific resort./Font was difficult to read./Not enough details about the activities offered./I did not encounter any problems or frustrations./Other: _____]

- What aspects of the Club Med site, if any, did you find particularly helpful or useful while looking for activities offered for a specific resort? [Text response]

- How difficult or easy was it to find activities offered for a specific resort on the Club Med site? [Rate 1 to 7; 1 = Extremely difficult, 7 = Extremely easy]

Case Study: Club Med and Beaches.com Competitive Analysis (continued)

- Overall, how would you rate your experience on the Club Med site?

	1	2	3	4	5	6	7	
Difficult to use the site								Easy to use the site
Not at all organized								Very organized
Frustrating experience								Enjoyable experience
Pages loaded slowly								Pages loaded quickly

- Now that you have spent more time exploring the site, how likely are you to:

	Definitely will **NOT**					Definitely **will**	
	1	2	3	4	5	6	7
Return to the Club Med site in the future							
Recommend the Club Med site to your friends and family							
Book a vacation on the Club Med site							

DATA GATHERING

The panel vendor sent study invitations to recruit participants, and the researcher monitored the completion rate. The study closed automatically when the desired number of completed studies was reached. We read the free-response comments to double-check the quality of responses.

ANALYSIS AND REPORTING

To analyze the data, I used WebEffective's online reporting tools, which provided views of the key metrics, task success rates, and browse times (see Figures 6.4 and 6.5). This included analyses of

- Qualitative and quantitative feedback associated with the tasks

- Behavioral data using clickstreams, which showed navigational paths; client actions, such as hovering, scrolling, clicking, and entering data into form fields; and behavioral data, such as total time on task, page load time, and number of pages visited

Case Study: Club Med and Beaches.com Competitive Analysis (continued)

- Attitudes, preferences, and motivations of target users

- Overall satisfaction, organization, and frustration for each site

Key Performance Indicators		
KPI	Club Med	Beaches
Average Success Rates	60.0%	67.5%
Average Browse Time	5.1 minutes	4.5 minutes
Average Page Views	9.3 pages	8.5 pages
Average Page Load Time	9.14 seconds	8.44 seconds
Average Time Spent on Evaluation	42.5 minutes	41.1 minutes

FIGURE 6.4
Key performance indicators, including success rate, browse time, and page views. The Beaches site unambiguously comes out on top here.

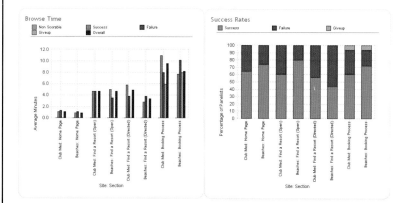

FIGURE 6.5
Browse times and success rates, visualized in graphs for ease of analysis.

I later provided recommendations for improving the customer experience, based on best practices and user feedback.

More users failed and had more difficulty with the directed task "Find a specific resort" than with the other tasks. Users spent the most time completing the booking process task.

Case Study: Club Med and Beaches.com Competitive Analysis (continued)

When looking for a resort, most users on the Club Med homepage started browsing by drilling immediately into a specific resort or by starting the booking process, whereas on Beaches they started browsing all resorts, thereby providing some initial high-level comparisons.

- 74% felt the Beaches homepage gave them a good understanding of what the site had to offer, compared to 64% on Club Med.

- 62% found the amount of information on the Beaches homepage appealing, compared to 38% on Club Med.

On the other hand, aspects of the Club Med homepage that users liked were the following:

- The photos: 70% found them appealing vs. 50% on Beaches.

- The special offers listed: 62% found them appealing vs. 36% on Beaches.

On the Club Med site, over half of the users first clicked on a specific village to begin looking for a resort, while most of the remaining users went to the booking tool or used the Village Finder (see Figure 6.6).

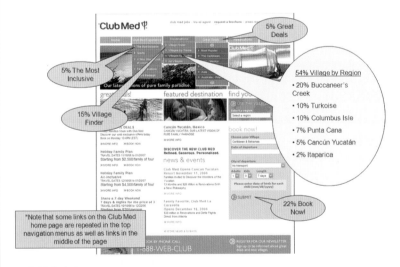

FIGURE 6.6
The majority of users began the task by clicking on a specific village on the Club Med site.

Case Study: Club Med and Beaches.com Competitive Analysis (continued)

In comparison, very few users went straight to a specific resort on the Beaches homepage; instead, many went to browse the whole collection first (see Figure 6.7).

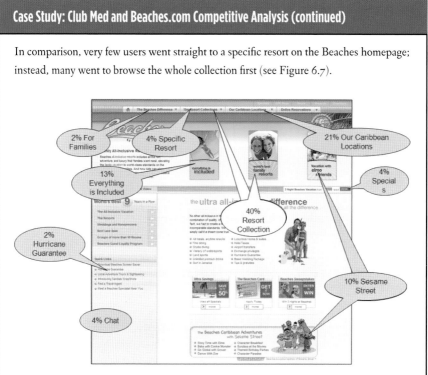

FIGURE 6.7
On the Beaches site, by contrast, people tended to click on the resort collections instead of specific villages.

Finally, while 65% of the Club Med visitors self-reported success, only 20% actually completed the booking process. The behavioral data collected by WebEffective while participants were interacting with the site showed us how many visitors dropped off at each stage (see Figure 6.8).

Some automated research tools, including WebEffective, provide the ability to trigger questions whenever a study participant drops off a defined process (e.g., shopping cart checkout, site registration, or loan application), which helps us understand *why* they dropped off. This trigger capability emulates moderated methods, where you'd be capturing users' thoughts within the context of the ongoing task.

Case Study: Club Med and Beaches.com Competitive Analysis (continued)

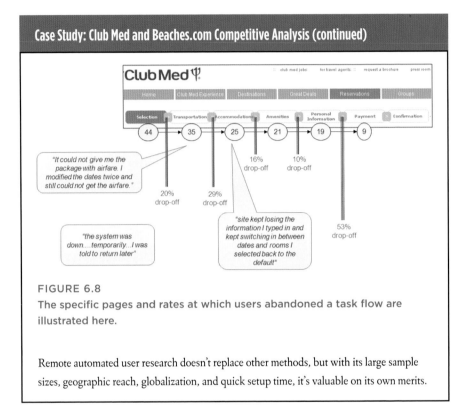

FIGURE 6.8
The specific pages and rates at which users abandoned a task flow are illustrated here.

Remote automated user research doesn't replace other methods, but with its large sample sizes, geographic reach, globalization, and quick setup time, it's valuable on its own merits.

Remote Card Sorting[1]

Card sorts have been around since long before computers. You give participants a stack of randomly shuffled cards, each with an item written on it, and then ask the participants to sort the cards into groups that make sense to them. The idea is to get a sense of how users naturally categorize different concepts so that you can structure your interface's categorization scheme correspondingly.

In contrast to task elicitation, card sorting occupies the conceptual end of the research spectrum but still traditionally falls into the category of UX

---

1   Thanks to Donna Spencer, author of *Card Sorting*, for her input in this radically condensed primer.

research. You perform a card sort when you want to understand how to categorize or organize a set of elements, an issue often related to the design of an interface's information architecture (IA). Some common examples of questions that can be addressed by card sorting include:

- What should the basic navigation categories for my Web site be?

- How should my site be structured?

- Which pages should I put these bits of content on?

- How should these bits of content be labeled or grouped on the page?

- Where in the site's navigation do these pages belong?

There are two kinds of card sorts: *closed sorts*, in which users sort the cards into categories that you've defined in advance, and *open sorts*, in which users come up with their own categories to sort the cards into. Closed sorts are useful when you have an existing IA or content categorization and need to figure out how to categorize a bunch of different elements. You can also use multiple sorts to compare different IA schemes. Open sorts are good for gathering insights about IA design or content categorization.

Open Sorts

The main challenge of designing an open sort is properly choosing the content you're trying to sort out. If you just throw in a bunch of items that have nothing to do with one another, the results will be all over the place. All the items need to be on what Donna Spencer calls the same "level" of content: they all need to be related in some very broad sense, like "potential categories" or "page names" or "things you can do with the interface."

We recommend not including any more than 50 items at the most, and even 30 can be a lot for complicated sorts. The longer the task, the greater the chance that your remote participants will get bored and distracted and either drop out or do a rush job. To keep the number down, include only the most representative items; if you're positive that Items B–E would be sorted the same way as Item A, you don't need to include Items B–E.

Items should be given self-explanatory names. The user shouldn't require any special knowledge to understand what these names mean. (Some tools, like

WebSort, will allow you to include a more detailed description of each item with a mouse rollover, but that'll make the sort take longer, and you should reduce the number of items to compensate. See Figure 6.9.) Make sure not to inadvertently create associations between items with the way you name the items. For example, if two of the items have the same brand names (ACME Googlydooter and ACME Figgisfiddis) or shared words (matchstick, matchlock) in the title, you might want to delete them to prevent the users from grouping them together based solely on similarity of title.

FIGURE 6.9
The online card sorting tool WebSort. Participants can simply drag cards onto the field on the right to create groups and add cards to them.

If your tool allows, it's good to set a minimum number of groups that the users are allowed to make to prevent lazy or smart-alecky users from separating items into two piles like "Good" or "Bad." It helps if you warn them against vague and lazy categories like this in the introduction to your study.

Closed Sorts

In addition to designing the cards, a closed sort will additionally require you to come up with the categories you're asking your users to sort into. In many cases this will be easy. If you're trying to sort content into a handful of existing pages, for example, you can just use the names of the pages as the category names. Sometimes your sections make for awkward card sorting category titles—for example, if you have the sections "Products," "Services," "About," and "Help," these titles may be slightly vague to users who aren't seeing them in the context of a Web site. In that case, you may want to inform the users beforehand what each category is meant to represent—in this case, sections on a Web site.

Moderated or Automated Card Sorts?

Even though most of the card sorting services out there (OptimalSort, WebSort) are set up for fully automated testing, nothing says you can't moderate your card sorts. It can be extremely useful to hear users talk through their card sorting tasks as they perform them so that you understand the users' thought processes for coming up with the categories. And you can always mix and match approaches; for instance, you could do 5 moderated and 50 automated card sorts, as we did in the following case study.

If you decide to go automated-only, be sure to encourage your users to make written comments so that you get as much of that valuable qualitative context as possible. Just like task elicitation, card sort tools usually provide some way for users to comment on their task either before or during the session.

Case Study: Product Category Card Sort

Fellow Bolt | Peters researcher **Alana Pechon** *provided the source material for this case study.*

In 2008 we conducted an automated card sort for a large communications hardware company, *Polycom*. We set out to determine where users would place a new product in the existing IA. The study consisted of a combination of both remote moderated and remote automated card sorting exercises.

To conduct the card sort, we used the Web card sorting tool OptimalSort. This study consisted of 5 moderated and 50 automated card sorting exercises, completed at the participants' own pace.

For client confidentiality purposes, we're not able to share the results in full detail; here we'll focus mainly on the method.

Case Study: Product Category Card Sort (continued)

RECRUITING

Participants were recruited using a short screener linked from several pages within the Polycom Web site. For the automated phase of the study, we offered visitors a 1-in-10 chance of receiving a $75 Amazon gift certificate in exchange for participating in the study, whereas all moderated users were offered the $75 Amazon gift certificate. The following data were gathered from each recruiting survey respondent: name, email address, job title, organization, size of their organization, affiliation with Polycom, reason for visiting the Web site, and Polycom product familiarity. From moderated participants, we also gathered phone numbers.

We invited respondents with an emphasis on prospective and current Polycom customers.

METHODS

The first phase of the study consisted of five one-on-one usability interviews, each lasting between 35 minutes and one hour. Testing was conducted remotely from Bolt | Peters headquarters in San Francisco, California. Users' screen movements and commentaries were digitally captured.

The second phase of the study was an automated open card sort exercise, completed online by 50 users. Testing was conducted remotely, with users participating from their own workstations and working at their own pace. Participants were given a selection of generalized product descriptions and asked to organize these descriptions into groups that made sense to them. They were also asked to comment on their task afterward. The raw results were then exported from OptimalSort and uploaded to the free online tool Card Sort Cluster Analysis Tool 1.0 to produce a similarity matrix, which measured how often cards were placed together and could be used to show averages of groupings (see Figure 6.10).

RESULTS

Overall, we found that users did not think about products in terms of what they could do, but rather how and where they would be able to use these products. They categorized products based on the contexts or scenarios in which the products would be used. Will this be used in an individual's office? In a conference room? Does this belong in a rack room or a communications closet? How many people will be using this at a time, and

Case Study: Product Category Card Sort (continued)

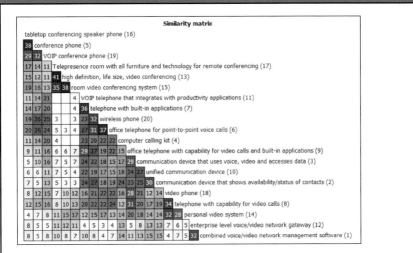

Similarity matrix

tabletop conferencing speaker phone (16)

38	conference phone (5)
29	32	VOIP conference phone (19)
17	14	11	Telepresence room with all furniture and technology for remote conferencing (17)
15	12	11	41	high definition, life size, video conferencing (13)
19	16	13	35	38	room video conferencing system (15)
11	14	21				4	VOIP telephone that integrates with productivity applications (11)
14	17	20				4	36	telephone with built-in applications (7)
19	26	25	3		3	23	32	wireless phone (20)
20	26	24	5	3	4	27	31	37	office telephone for point-to-point voice calls (6)
11	14	20	4			23	20	22	22	computer calling kit (4)
9	11	16	6	6	7	29	27	19	22	15	office telephone with capability for video calls and built-in applications (9)
5	10	16	7	5	7	24	22	18	15	17	29	communication device that uses voice, video and accesses data (3)
6	6	11	7	5	4	22	19	17	15	18	24	27	unified communication device (10)
7	5	13	5	3	3	24	27	18	19	24	23	25	30	communication device that shows availability/status of contacts (2)
8	12	15	7	10	12	16	21	22	22	16	28	21	12	14	video phone (18)
12	15	16	8	10	13	20	22	22	24	12	31	20	17	19	34	telephone with capability for video calls (8)
4	7	8	11	15	17	12	15	13	14	20	18	14	14	32	28	personal video system (14)
8	5	5	11	12	11	4	5	3	4	13	5	8	13	13	7	6	5	enterprise level voice/video network gateway (12)
8	5	8	10	8	7	10	8	4	7	14	11	13	15	15	4	7	5	32	combined voice/video network management software (1)

FIGURE 6.10

Similarity matrix generated by Card Sort Cluster Analysis Tool, which illustrates the affinity groups into which participants sorted the cards.

where in our office will they be? As more than one user explained during the moderated portion of the study, when they search for a product, their initial thoughts about the product were not organized around the technology it employed, but who would use it. This point of view was reflected over and over in the categorizations, as users organized their groups not according to the feature set or technology of a product, but the context in which they imagine it will be used. Several open-ended comments also illustrated this ambivalence.

Being extremely accustomed to the existing Polycom IA, users accordingly titled their categories similarly to the existing IA. However, the items in the various groups did *not* follow the Web site's classification scheme, and in some cases the items contradicted their labels and conformed to the "context categorization" scheme described earlier, in spite of the labeling.

Based on these results, we recommended that in the long run the IA should be overhauled to reflect the products' manner of usage rather than their functionality, following the categories we uncovered in the card sort.

Surveys

We would be remiss if we didn't at least mention surveys, which are a huge part of the research world; however, it's market research, not UX research. Surveys can ask abstract or concrete questions, but they are strictly nonbehavioral, unless you happen to be curious about the way people interact with surveys.

Why do people do survey studies? One reason is that they're so easy. Designing questions is usually as easy as writing a few sentences. If you're recruiting from your Web site using a screener, you can just insert the questions directly into the screener, bypassing the entire step of sending invites to prospective participants. And surveys are great for getting a quick read on user preferences ("What do you think of the colors on the site?"), demographics ("How many rabbits do you own?"), psychographics ("What do you want to accomplish in life?"), or new ideas for your interface, which are more reliable now that you're working with a larger sample size.

We personally like to mix in a few survey questions into our automated behavioral studies, just to segment and cross-tabulate the users (although you could also accomplish that through the recruiting screener) and to be able to have something to say about both user behaviors and opinions. But we're not gonna lie and pretend that we're experts on designing surveys; we're UX researchers.

One very important thing we can urge you *not* to do, however, is rely on surveys to address issues that should be addressed in a behavioral study. Don't ask, for instance:

- "Where did you have the most trouble using the site?"
- "How would you improve the login process?"
- "Would you describe the site as 'user-friendly'?"

If you do ask such questions, you will fail. Instead, add a task-elicitation component to the study to get behavioral feedback and use opinion-based stuff only to shed light on the behavioral findings. What users want and what they *think* they want are two very separate things. Don't try to figure out how to improve the functionality and interaction design of your site by just asking users what they think. We're begging you.

Other Methods

Some of the other automated research methods out there are:

- **Web analytics.** While monitoring your Web traffic and stats is technically a form of automated research, it's extremely blunt and usually can't be used to address specific design questions with any clarity. At any rate, it's obviously a good idea to use services like Google Analytics to answer certain broad questions about your site, such as what sites users are coming from and where they're going, where they arrive on your page, where they leave, what parts of your site are most popular, how your traffic has progressed over time, and so on. Good luck trying to use analytics to answer anything more specific than that.

- **Multivariate testing.** Multivariate testing services work with your Web site's content to serve different versions of the same page to visitors to determine which version performs best. (Back in the day, when it was called "A/B testing," people offered only two versions of the Web site, but researchers discovered that two wasn't enough for useful feedback.) Multivariate testing services usually offer ways to parcel your content and design into 5–10 blocks, serving up different combinations of those blocks and then running an automated test along a set of predetermined metrics (usually clickthrough rates).

- **Online diaries and ethnographies.** Services like QualVu and Revelation are offering tools that enable users to maintain regular diaries (both written and video-based) about their experiences with a particular brand, product, or interface. These are generally used for market research, but there's no reason you couldn't use these tools to prompt users to interact with a product in front of a camera. The downside is that these interactions would be entirely (and by definition) automated, which makes it more likely users may veer off into giving opinion feedback, and makes it impossible to probe further on interesting issues as they come up.

- **Annotation.** Annotation tools allow users to leave feedback by placing written notes on a static image of the interface, allowing them to comment directly on a design. This has most often been used for "concept testing," a market research approach. Some variations of annotation, like the now-defunct MindCanvas's "Sticky" tool, also allow you to design the notes yourself, constraining the users to a limited set of possible comments: "This is good," "This is confusing," "I hate this," etc.

- **Input tracking.** Input tracking refers to tools that make recordings of user mouse movements and keyboard input. This approach is sort of like a zoomed-in version of analytics, in the sense that you can get a very close and exact record of what a user is doing on the site, but without any specificity about task context, user intent, or who the user is. In combination with a task elicitation tool, the input tracking could be informative, but since most input tracking tools are anonymized, there's currently no easy way to correlate the input tracking results with the task elicitation results.

- **Game-based elicitation.** Not sure if anyone uses this approach anymore; it was a big part of MindCanvas's methods. (By the way, yes, we miss MindCanvas.) "Game-based elicitation" is a blanket term for any tool that takes a simple conceptual exercise and turns it into a task with artificial constraints or rules to get users more engaged. Examples include "Divide the Dollar," in which users divide a limited amount of pretend money among a set of items of their choosing as a way of indicating their preference for them, and Revelation's "Help Mabel," in which users answer questions posed by an imaginary grandmother named Mabel. We haven't been able to find much evidence that you necessarily get better results just by putting things into a game, but for those of you who are bored with Likert scales, there it is.

Case Study: Products or Services?

In 2007 we conducted an automated study for a large software developer, for which we set out to determine whether users preferred a site structure that allowed them to navigate the site by products first or by services first (i.e., Tech Support, Training, Downloads). The study consisted of a combination of card sorting, task elicitation, and survey exercises.

To conduct the card sort, we used the now-defunct automated research tool MindCanvas by Uzanto (though other task elicitation tools could have been used). This study consisted of 107 automated usability exercises, completed at the participants' own pace.

For client confidentiality purposes, some elements of the study have been anonymized, and we weren't able to include screenshots of the pages being tested. Product names are referred to as [X], [Y], and [Z], while the client is referred to as [Client].

RECRUITING

Participants were recruited using a short screener linked from several pages within the client's Web site. We offered visitors a 1-in-10 chance of receiving a $75 Amazon gift certificate in exchange for participating in the study. The following data were gathered from each recruiting survey respondent: name, email address, reason for visiting, client's products owned/used, and job title/industry. We invited respondents based on which of the client's products they owned, seeking to get a good mix of different products.

METHODS

Users were shown an image of the page and asked to click in response to the following tasks.

1. "Where would you go to find Web support for your [X] product?"

2. "Where would you like to OR expect to find Web support for your [X] product?"

3. "Where would you go to find Web support for your [X] product?"

4. "Where would you like to OR expect to find Web support for a non-[X] [Client] product, such as [Y]?"

Case Study: Products or Services? (continued)

The users were then asked to sort the following nine randomly organized cards into groups that made sense to them:

- Technical Support for [X]

- Training for [X]

- Downloads for [X]

- Technical Support for [Y]

- Training for [Y]

- Downloads for [Y]

- Technical Support for [Z]

- Training for [Z]

- Downloads for [Z]

After that, users were shown images of two different pages within the site, which served roughly the same function but had different organization (product-oriented vs. services-oriented) and were asked the following:

- "Click on the page you usually visit to find support for your product."

- "Click on the page you think makes it easier to find support for your product."

The users were prompted to leave additional open-ended comments after completing this A/B decision. They were then thanked for their participation and told that they would be notified if they were selected to receive the incentive.

RESULTS

"Where would you go to find Web support for your [X] product?"

- [X] Product links   85%

- Technical Support   9%

Case Study: Products or Services? (continued)

- Global Support link   3%

- Other   3%

"Where would you like to OR expect to find Web support for a non-[X] [Client] product, such as [Y]?"

- Technical Support   26%

- [X] Product links   15%

- Other   15%

- Home link   12%

- [redacted]   10%

- Global Support link   7%

- Subscription Home   6%

- Training link   6%

- Subscription Help   3%

"Click on the page you usually visit to find support for your product."

- Page A   62%

- Page B   38%

"Click on the page you think makes it easier to find support for your product."

- Page A   66%

- Page B   34%

Where Do I Learn About Tools?

In contrast with moderated techniques, which are more or less the same no matter what screen sharing tool you use, the functionality of the automated research tools you use has a huge impact on the type of info you're able to gather. Will you be able to conduct competitive research? Can you test prototypes? Can you export your findings? How will the results be visualized? Does it come in different languages? Can you combine different methods using the same tool? Are there integrated recruiting tools? What will it cost?

In this chapter, we tried to be as general as possible about our methods to suit a wide range of tools, but sooner or later you're going to have to choose one. Since tools come and go very quickly, we've decided to bracket off all the information about research tools into its own chapter. If you're looking to learn about the different tools (currently) available, skip ahead to Chapter 8, or you can always check http://remoteusability.com for the latest. Otherwise, read on.

Chapter Summary

- Automated research is effective for gathering large-scale, quantitative feedback to address targeted usability issues.

- Automated methods are not a cheap replacement for moderated methods; they have totally different purposes.

- Live recruiting methods still work, but you will have to do more of it, and send out several rounds of invitations. When recruiting for an automated study, it's standard to offer lower incentives.

- Task elicitation is the most prevalent form of automated usability research. Remote card sorting is good for solving organization and categorization issues. Surveys are mostly an opinion-based research format and don't capture behavior well.

- When possible, gather both quantitative and qualitative data in the same study.

- Your choice of tool or service matters a lot; see Chapter 8 for a guide to currently existing tools.

CHAPTER 7

Analysis and Reporting

A s we've mentioned, one great thing about remote research is its ability to capture people doing real tasks on their actual home or office computers. And the ease and low cost of recording the research sessions digitally allow you to use video editing software for lots of purposes. These aspects of remote research can help streamline analysis, make new observations based on the users' native computing environment, and create a more complete picture of the people who use your product.

In this chapter, we'll focus on what's new and what's different when analyzing and presenting the findings of remote studies, both moderated and automated.

Moderated Analysis

The core task of analyzing and collecting data from a remote moderated study isn't much different from an ordinary in-person study; you're still trying to observe user behavior. The differences have mostly to do with what you can glean from the users' native environments and how you can best extract and synthesize all the recorded data.

The Technological Ecosystem

We strongly recommend talking to users who are using the computers they'd be using for precisely the tasks you're testing. It gives you the opportunity to observe the users' computer environment—all the programs, open windows, layout, Web sites, and other stuff happening on the desktop—which we like to call the *technological ecosystem* (see Figure 7.1). For you ethnographers and UX practitioners out there, this is the remote research equivalent of "usage context," and it's important for the same reason: it gives you a detailed look at how people organize and use different Web sites, programs, and files together to get things done. As Brian Beaver mentioned in Chapter 1, tasks like online comparison-shopping are rarely performed in a vacuum. Web browser tabs, notes, and physical brochures may be used in conjunction with any given commercial Web site to help users compare and research products.

FIGURE 7.1
Here is a typical user desktop. Multiple open windows, tabs, shortcuts, taskbar items, and desktop files are discernable at a glance.

Insights may come from anywhere on the users' screens, but a handful of areas deserve especially close attention because they're often related to tasks that users perform on their computers: Web bookmarks, the desktop, and the Web browser configuration. First, the bookmarks: how many are there? How are they organized? Are they organized? Do users have bookmarks for Web sites similar or related to your interface? You can have the users explain these details to you, as well as which bookmarks they visit most frequently, and what purposes they use them for. You could even ask users to visit some relevant bookmarks, if time allows. You should also look to bookmarks for insights on information architecture. Users frequently bookmark pages deep within a site, which can indicate that content there might be important to investigate in later analysis. (It could also mean they don't know how to use bookmarks.)

Correspondingly, desktops are a great way of seeing what users do offline. You can see the shortcuts, program links, and files they've stowed for easy access. You can get hints about their information gathering and organizing habits. Is their desktop hopelessly cluttered or spotlessly tidy? Are there particular types of files lying around—PDFs, Word documents, shortcuts? The files are also sometimes clues to the types of files they use or tasks that they perform most often.

Even the look and feel of the desktop can offer unexpected human touches. Desktop wallpapers and themes, while perhaps not helpful for understanding behavior, can at least give you the sense that you're talking to a real live person, and promoting empathy is usually a good thing.

Particularly with the rise of customizable browsers like Firefox, the Web browser has become another point of interest. Often, when recruiting users live, you'll find that they have a few Web sites already open. If they haven't already closed them down for privacy's sake (as you instructed earlier), you should pay attention to their relation to the users' tasks, whether or not (and how often) the users switch between them, and which tabs get the majority of their attention. It's often tough for Web designers to anticipate how their site will be used with other sites. In Brian's example in Chapter 1, the observation that users heavily relied on third-party review sites alongside the SonyStyle site to evaluate products holistically was a major finding.

You may find that your users have restrictions and quirks about their computer usage as a result of their computing environment. Most commonly, you'll find that computers at corporate institutions have limitations on email, IM, Web access, downloading, and software installation, and it's good to know exactly how users work around these limitations (as long as it's allowed, and you can still get screen sharing up and running!). Financial and government institutions often have loads of security restrictions, while medical institutions may have lots of safeguards in place to enforce confidentiality of patient information. Particularly while testing internationally, you'll want to see if Internet connection speed affects what kinds of sites users are able to visit and whether they're using it in a public space (e.g., the "PC Bang" Internet cafes that are popular in Asia). Paying attention to computer restrictions is of particular interest for studies of niche user groups, especially if such restrictions are an issue that will affect a broad segment of the interface's intended audience.

Aside from the stuff that's on the participants' computers, you can get a hint of the physical environments and artifacts that people use along with their computers sometimes. Hundreds of users have described to us the notebooks, binders, and PDAs they use to store their dozens of login usernames and passwords, and some have told us about the way they print out information (bills, receipts, directions, travel plans, personal health research, etc.) for later reference. In one study for a car insurance Web site, we had one user physically get up from his desk in the middle of the session to check the vehicle identification number (VIN) on his truck, which he needed in order to fill out a form on the site.

NOTE THINGS WE'VE NOTICED BEFORE

Following are some things we've noticed in people's technological ecosystems that have led to interesting discussions and insights:

Ingenious methods of storing login names and passwords. Some users kept them on a piece of paper beside their desk; others had them stored in Word documents; others wrote them out in their bookmarks, right next to the Web URL.

Offers to take us on a guided tour of the research libraries in their My Documents folder on their laptops, which they preferred over using the company intranet or public drives. Seeing user-developed taxonomies showed us how users would label items on the site if they had the chance.

Screen resolutions. A group of graphic designers using design software had set their screen resolutions extremely high, a very important consideration in interface design.

Plug-in usage. Online shoppers—who were not otherwise technologically savvy—had the latest Firefox plug-ins for automatically filling out e-commerce forms.

To-do lists. During one interview, a mother looking up information to care for her sick child referred to a to-do list on her desktop, which ended up revealing to us how the Web site she was visiting was just one component of a complex workflow of managing phone calls, emails, and appointments.

Making Highlight Clips

Making highlight clips of research sessions is a road-tested and effective way to show off your findings. You take 5- to 20-second snippets of video from each of your research sessions and compile them into a 3- or 4-minute video that illustrates a theme. Most people use basic programs like iMovie or Windows Movie Maker for this; we've found Sony Vegas to be a good (though not flawless) tool. There's also Final Cut and Adobe Premiere for the savvier moviemakers.

We're not going to teach you how to use each of those tools; that's crazy talk. Just spend some time in the tutorial. Mostly, it should just amount to dragging video segments into a timeline and cropping them. A few tips about editing videos for remote research, however:

- **Be conscious of your editing choices.** Since most of what you'll see in a session recording is the user's desktop, which is static much of the time, it's possible to cut out superfluous gaps and pauses in the video without making it appear jumpy, as it would if you edited out footage of a live person. That makes it very easy to mislead people, which you don't want to do. Be careful that your edits don't affect the meaning and integrity of the video; for example, if cutting out parts of a clip make it appear that the user is having a much easier time with a task than he/she actually did, then don't do it, or use a video effect (like a fade-out/fade-in, or cross-dissolve) to make the edit clearly apparent.

- **Sometimes video recordings of your sessions come out damaged, corrupted, or otherwise unplayable.** This usually happens when the computer processor gets overburdened while the video file is being created. Sometimes there will be no video or audio, or it won't open at all. The best way to prevent this problem is to avoid doing too many things on your computer while the video is being rendered out at the end of your session and to make sure your computer has plenty of RAM so that opening other programs will interfere less with the render process. If you end up with a damaged video anyway, there are a couple of steps you can take. First, try opening the video in different media players. VLC is a versatile, open source player, and RealPlayer covers some of VLC's blind spots

(RealMedia files, FLV files). Video editing programs like Sony Vegas and Final Cut also have good compatibility. If that doesn't work, find a video converter program that can take one video file and convert it to another. And remember to back up the videos before repairing them.

- Pixellate, blur, or mask any private information that appears onscreen, using your video editing software's "mask" tools. This information includes users' credit card information, home addresses, email addresses, and so on. This will require some knowledge of Final Cut Pro or iMovie '09.

- If parts of the conversation are distorted or garbled, make subtitles. Make sure the subtitles don't block any portion of the screen that's important for understanding what's happening.

Automated Analysis

The way you visualize and analyze the findings has mostly to do with the tools provided by the online automated service you use. We'll cover many of these tools in the next chapter.

What's important to remember is that automated research doesn't necessarily have to involve any high-level statistical analysis. Large-scale quantitative data are attractive to number crunchers, but we want to caution you again that "science-y" research won't necessarily inspire new designs or help you refine and improve your interface. In most cases, automated analysis comes down to a common-sense comparison of numbers, followed by a lot of hard, exhaustive thinking about *why* the numbers came out that way.

Here we'll get into methods of digging through your data to get the basic insights that will add a quantitative dimension to your qualitative findings.

Task Elicitation and Analytics

Three popular metrics for evaluating quantitative user performance on a task are success rate, time on task (i.e., how long users take to complete the task), and the "first click"—where different users clicked to perform a particular task. All three metrics address the fundamental question of whether a particular task is intuitive, and with a well-crafted question, you can usually get an answer you're looking for just by glancing at the report generated by your automated research tool (see Figures 7.2 and 7.3). Any automated tool worth its salt will provide a simple way of presenting these metrics in a graph, spreadsheet, or clickmap (an image of the interface overlaid with a graphical layer that indicates where users clicked).

FIGURE 7.2
WebEffective's "Key Performance Indicators," including average success rates, time-on-task, and page views.

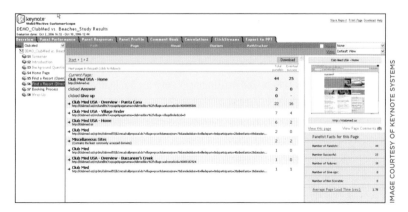

FIGURE 7.3
WebEffective participant success rates on various tasks.

Other interesting metrics to examine include the most common paths taken to accomplish a task, back button usage (if the automated tool can track it), page views, and the drop-off rate for a particular task (what proportion of users abandon the task and at what point). About the drop-off rate: be careful that abandoning the task isn't confused with abandoning the study. Some tools keep track of this, but some don't, so just make sure that the results don't include abandoned studies.

You can get deeper into the data by using a spreadsheet's basic functions, if the automated tool doesn't handle it for you already. There's cross-tabulation, or evaluating one metric in light of another: "Of all users who clicked on the login button first, 80% were able to successfully log in." You can also take the average and standard deviation of certain metrics for purposes of synthesis, but be careful that this doesn't obscure any of the subtle points about the data. If you note that the average time on task is 2 minutes, make sure that this is consistent across all relevant user segments as well. (It might take most people 1.5 minutes, but older users 6 minutes.)

You shouldn't necessarily limit yourself just to the data you've collected for one particular study. If you are doing a follow-up study on an interface you've tested before, consider including the same tasks as previous studies, to form a basis of comparison. If the success rate on a task differs significantly from the success rate on the same task in a previous study, you should obviously look for the reason.

Qualitative feedback can also be used to indicate and define the themes that come out of your research. By loading open-ended responses into a spreadsheet or word document and running them through a word analyzer, which displays the most frequently occurring words, you can get a rough idea of some of the issues that are on users' minds. Some tools have built-in functionality that does this for you. If more than half of the users address a particular issue ("navigation," "search," etc.), you're probably on to something major; even a quarter will probably still be pretty interesting. Due to typos, the tally won't be exact, but qualitative data can be used to guide your insight into user motivations. The same goes for the closed-ended subjective task ratings: they're opinion-based but can point toward interesting behaviors.

And how about Web analytics? Without the use of explicit task elicitation, analytics are trickier to use because there's no explicit context for user behaviors. You can say, "90% of all visitors click on the green button," but you can never be certain *why*. The greatest strength of analytics tools is that they track information persistently so that you can see how traffic evolves over a span of time; even less than task elicitation methods, though, you have no idea why anything is happening. We encourage using analytics tools for monitoring high-level things like traffic volume, where your traffic is coming from (in terms of both physical location and Web referral), what pages users land on, what pages are most popular, how long users are on your site, and popular search terms. But you should stop short of drawing assumptions about the users' motivations and performance based solely on analytics data. On the other hand, analytics combined with task elicitation can guide you to where you should pay attention. If there's a discrepancy between your study findings and the Web site's analytics ("80% of study participants clicked on the green button, but only 40% of our general Web audience does"), it could mean that the task design was flawed, the target audience of the study differs from that of the main audience, or that there's an unforeseen issue altogether.

Card Sorting[1]

As mentioned in Chapter 6, card sorting comes in two major varieties: open sort and closed sort. Closed sorts are pretty straightforward and only require you to break down which categories were most favored by the users. With multiple user segments, you'll obviously want to identify where the segments sorted differently. See Figure 7.4 for sample card sort results.

Open sorts take longer to analyze and interpret because users create their own categories, and the degree of consistency in open sort categories can vary between users and with larger groups. There are tools to help perform automatic cluster analysis on card sorts (see the card sort case study in Chapter 6), but you may have to analyze by hand. If you start with a clear

---

[1] Some of the following discussion is informed by discussions with Donna Spencer and her book, *Card Sorting* (Rosenfeld Media, 2009).

idea of how to analyze and group the findings, though, your analysis shouldn't take longer than a few hours of tedious data wrangling—fewer if you drink a few Red Bulls first.

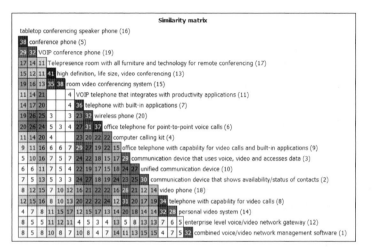

FIGURE 7.4
Card sort results placed into a similarity matrix with the Card Sort Cluster Analysis Tool.

First, load the raw results into a spreadsheet; most online tools will allow you to export the results to Excel. You'll be performing a kind of meta-card-sort on your card sort results. Go through the first few categories, grouping related or redundant categories into higher-order ones. For example, let's say one of the cards is "Hammer": if one user categorizes it as "Tools," another user as "Appliances," and a third as "Hardware," you might just label them all as "Tools." From then on, you'd be able to sort anything that fits into any of those three categories as "Tools," adjusting or renaming the category as you go along.

Be flexible. If you see that someone categorized "Fork" as a "Tool" later, you might want to broaden the category to "Implements." Or you might see a more reasonable categorization scheme further down the line, which means you'll have to backtrack a bit. If you create a quick numerical code for each category ("1–Tools," "2–Appliances," "3–Decorations," etc.), you can quickly go down the spreadsheet and tag each entry with a number.

If you're using a tool that allows for open-ended user comments (which we do recommend), it's good to skim through at least a few dozen to see if there's any helpful information that might help you refine the categories. When all's said and done, you should have a rough sense of what the major categories are, and each item in the spreadsheet will be coded in one of the categories. At this point, using the categories you've made, you'll be able to treat the results just like a closed sort. It's good to get someone else to verify your groupings, just to be thorough.

Interpreting card sorts, especially open sorts, is inherently subjective and will require you to draw on common sense and UX experience and know-how. Donna Spencer describes a study in which most users created about 20 categories on average; the designers took this result to mean that their IA should have 20 navigation categories. Naturally, you, as a shrewd UX practitioner, would know that navigations should generally be compact, so you'd collapse those 20 categories into more general categories.

As before, we refer you to Donna Spencer's book, *Card Sorting*, for anything more complicated than this.

Surveys

So, here's the thing. Surveys are practically by definition used to capture self-reported user opinions and preferences, not behaviors, which means that they fall into the realm of market research and not UX research. To the extent that surveys can help provide context to other research findings, they can be interesting sources of insight, but parsing large-scale self-reported opinion data is beyond the scope of this book—and so, we move on.

Reporting

After you complete your research sessions and analysis, there are steps you can take before, during, and after the formal presentation of the findings to make sure your remote research findings are put to good use. If you took the time and effort to conduct user research on your own behalf to learn what users want out of your interface, then we salute you. You're in a good position to draw insight and inspiration from your research, since you saw it all firsthand. But if you're conducting your research on behalf of other people, you have the challenge of articulating your findings to people who may or may not understand, care about, or want to hear them. These people may look just like your otherwise-friendly colleagues, bosses, and clients, but don't be fooled! Research findings can be hijacked in a million ways, and it's up to you to stop that from happening.

Information by itself accomplishes nothing. If you don't organize, present, and follow through on the findings of your research, making sure that the right people hear and understand them, you run the risk of being ignored, and you might as well never have conducted the research at all. The right videos and presentation, on the other hand, can inform and inspire your audience, which is what it takes to really get things done. The usual report-plus-hour-long-presentation is simply not that effective. Think of it this way: if you're handed a brochure plainly stating a few ways to eat better and exercise more, there's a *chance* you might follow them, but you're not going to get fired up to do it. Most likely you'll think, "Well, duh," and then do nothing. But when you read a story about someone like you who made a few simple adjustments to his/her life to eat better and exercise more, you might say to yourself, "Yeah, I can do that."

In the following section, we'll focus on what you can offer besides the typical bullet-point report to communicate your findings usefully.

Before the Presentation

The last thing you want to do is present to a roomful of people who are hearing about the research project for the first time. You'll spend minute after dreadful minute scrambling to explain basic things about the methodology that should have been explained in advance. The easiest way to prevent this situation is to keep everyone in the loop as the research is ongoing, so they can become familiar with what the study is about and why you're doing it.

It's good to send out summaries of each day's testing results to any and all stakeholders who are willing to read them, with the caveat that the testing is still in progress, and findings may change later. The summaries may contain the following information:

- The number of users tested that day

- The number of users left to be tested

- A few stand-out quotes and comments

- Possible themes or areas of interest indicated by the research so far

- The remaining testing schedule (to entice more people to observe sessions)

- Any updates or information relating to the testing procedure

Additionally, if you have raw videos of the sessions and it's okay to make them available to the stakeholders (i.e., the videos don't need to be edited for privacy reasons, etc.), go ahead and share them. We find that linking to the raw videos from a Web page is the easiest way to do that, although file delivery services (RapidShare, YouSendIt, MediaFire, MegaUpload, NetLoad, FileFactory, etc.), and FTP can work as well.

If you have the time, sending along drafts and early versions of your report as they're completed to allow attendees to raise questions and concerns early can be good for getting a reading on what the stakeholders' concerns are in advance. Do everything you can to keep from getting mired in small quarrels during the presentation.

Presenting Moderated Findings

One of your challenges will be to convey the findings to the people in such a way that your users are real people and not faceless abstractions. When you're dealing with a set of users whose faces you can't see, it's easy to abstract them—"User #3 did this, User #7 did that"—but, of course, this isn't what you want. Aside from the standard highlight clips and quotes, there are lots of little ways to keep the emphasis on the users while presenting your findings. Having an image to pair with each user helps enormously. At the end of the research sessions (or in a follow-up email after the studies have concluded), consider asking users for pictures. Better yet, you can even request that the users take their own pictures in their computing environment, or along with the interface being tested.

In a recent study we conducted for a major consumer electronics manufacturer, we had our participants send us pictures of themselves along with the piece of hardware they were looking to replace. Lots of users have webcams or digital cameras and are more than happy to oblige. Keep in mind that the usual privacy and consent rules apply here (see Chapter 4). The pictures can be inserted into the highlight videos in a picture-in-picture or side-by-side format so that the images are juxtaposed with the behaviors (see Figure 7.5).

FIGURE 7.5
A side-by-side screenshot and user photo from our 2009 Wikipedia study.

Besides making each user stand out, you can also try to visualize the "big picture" ideas that define the themes of the study. Nothing does this better than well-edited highlight clips, but that's not necessarily your only option. With all the data you collect throughout a remote research project, there are many possibilities for slicing and dicing your information. Tagging users with various behaviors and habits can have two good purposes: it can be a simple way to categorize and characterize the users with just a few keywords, and it can also be aggregated to produce tag clouds using online tag cloud generators (Wordle, the ManyEyes double word tool, TagCrowd), which can visually highlight heavily recurring themes and give a quick at-a-glance overview of what the study was about (see Figure 7.6). You can also try seeing what comes up when you run your verbatim notes through there. It can make a fun slide or endnote to your presentation. Even the data from the live recruiting extract can be illuminating, providing a snapshot of the users who filled out your screener.

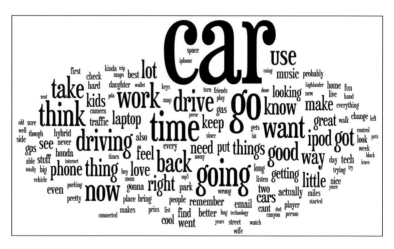

FIGURE 7.6
A tag cloud generated from our 2008 automotive study (see Chapter 9).

Presenting Automated Findings

Even if you design the perfect study, you don't want your findings to rely solely on the results of your automated research because you run a little risk of insufficient context. There are always a hundred good reasons why users chose to click a link, abandon a task, or sort a card into a category, and even their self-reported comments may not tell the whole story.

Attendees are likely to be curious about the specific numbers, so those should be part of the whole package of deliverables. Quantitative data are good for lending support to an argument, but you should not always be tempted to rely on graphs and charts to make your biggest points. You should present automated findings in conjunction with other corroborating (or contradictory) qualitative data. So if you have a chart demonstrating that 70% of users tended to click on the login button first, also include some qualitative behavior or feedback that sheds light on this behavior (e.g., video of users trying and failing to click on other things first, or comments like "I thought I wouldn't be able to do anything on the site until I logged in").

NOTE PRESENTING REMOTELY

Many of the tools that you use to see users' screens and communicate with them (like GoToMeeting) are actually designed for remote presentation and collaboration purposes, so naturally, you can use them to present your study findings. This capability is particularly useful if the reason you're doing remote research is to involve stakeholders who are geographically distant. With most current screen sharing solutions, it's difficult to show videos to observers, so instead of playing the highlight clips on your own screen, have the observers set up the highlight clips on their own computers, to play on your cue. (This is much easier if all the observers are in the same room and can watch the videos on a projection or large screen.)

Other Approaches to Presenting Findings

Card sorting expert Donna Spencer mentions that rather than presenting findings and then making recommendations based on those findings, she likes to do things the reverse way. With a dry erase board, she'll focus on sketching out and talking through possible design solutions step by step, citing the study results as needed to justify various decisions. This way, she says, you can focus on the design process itself, and synthesize a lot of the information, instead of spending time translating raw data and speculating about context.

UX luminary Jared Spool forgoes recommendations altogether. Instead, he'll present the findings and then allow the meeting attendees to draw their own inferences as he moderates the discussion to prevent it from getting bogged down in false or misguided assumptions. The idea there is to allow people to arrive at their own conclusions rather than be *told* that "X-finding requires Y-redesign," which increases their understanding and conviction in the research.

Our honorable foreword author (and president of Adaptive Path) Peter Merholz says that design is the best presentation for research data. He believes that if you're not creating concrete design ideas as part of conducting the research, and presenting those design directions, sketches, and concepts at the end of the research, you're simply wasting your time.

For typical studies, we usually take a middle approach between these three. We like to give a detailed walkthrough of the themes and recurring user behaviors of the study, with the major ideas illustrated by highlight video clips and simple, clean PowerPoint slides. These slides and clips are usually summarized with a user quote or short title that makes the problem relatable to anyone: "Users feel stupid when they try to edit the page," or "Nobody knows what the widget does." We also compile a more detailed document of findings, quotes, and usability issues, to appease developers and others who want a long list of details and minutia to mull over later.

Naturally, the details of this approach depend on the type of project we've been hired to do. If it's a large-scale, high-risk project, we document things more exhaustively. If it's a low-budget usability review, we stick to quick bullet points and go light on videos. If it's research that's intended to inspire a design, we'll forgo specific recommendations and instead produce pen-and-ink design sketches, photographs, longer videos, and presentations that attempt to tell a complete story about what roles the interface plays in users' lives.

Clearly, there's an endless number of ways to present findings. Whichever you take, you should have in mind the general principles you're trying to communicate. A humble usability study on a certain widget or element in your interface may not require much thought, but if you're testing anything of significant complexity, you want to get at how that part of the interface, and the problems and behaviors you've uncovered, relate to your big-picture goals as a team/company/country/basement tinkerer in Montana.

Fielding Questions from Newbies

Sometimes—there's really no stopping it—you'll find yourself in the position of having to field a lot of questions from people either unfamiliar with, or resolutely skeptical about, remote research methods. Here are some of the most common questions we've gotten during our research presentations, and our preferred method of dealing with them.

How can we tell how users feel about the interface if we can't see their faces?

First of all, when it comes to evaluating the usability issues in an interface, feelings don't actually mean a whole lot; user research is about behaviors, about what people can and can't do, and the techniques and workarounds they use to get things done. Since the major goal of the study is to examine those behaviors, it's sufficient (and, actually, a lot less distracting) to focus on users' onscreen behaviors. (That said, a fair amount of emotion comes through the users' voices anyway, so you actually can tell what they're feeling most of the time.)

Why didn't you ask the user to [click on X button/visit X page/use X feature] right away?

A big part of remote studies is about observing the users' natural behavior, and we wanted to see if users would [do X] on their own; usually, we only ask users to [do X] if they've overlooked it entirely.

The user said that she liked [X feature of the Web site], so why are you recommending that we change [X feature]?

The goals of this study were to derive insights from the ways that users behaved on the site, rather than their self-stated opinions of the site. Besides, many other users struggled with this feature.

How do we know that the users gave honest feedback, since they knew they were talking to someone involved with the interface?

You're right to be concerned. There's a well-known concept called the Hawthorne effect, in which study participants' behavior and feedback change under observation. However, that is part of any user research, especially in person. At the beginning of the session, the moderator indicates that he/she is not involved with the design, and that the users can be completely candid. We don't believe that this concern makes a significant difference in people's interface interactions.

Fielding Questions from Newbies (continued)

Can we follow up with the users after we've made the changes to the site to see whether they're satisfied?

That's possible, though it must be kept in mind that expressions of "satisfaction" are opinions, and the users will already have had experience with the interface, so they will naturally have less trouble and are likely to report greater satisfaction. A better approach would be to conduct another study with the same goals after the changes are made, to see whether user performance changes.

Following Up

After the curtain falls on the study, you have one more thing to do: make the findings as easy as possible for the designers and decision makers to refer to as they begin the redesign process. By putting up a Web site from which clients can quickly and easily download the project materials at their leisure, you can make it easy for people to access your findings and to share these findings with other people on their team or within their company, which can only help your cause. If you have some heavy-duty user consent to share the session videos publicly, you could even post them to sites like YouTube or Vimeo (as was the case for a study we did for Wikipedia in early 2009; see Figure 7.7), thereby adding to a body of knowledge that can benefit UX researchers all over the dang place. (That's a big privacy issue, though, so be sure you're playing by all the applicable privacy and consent rules.)

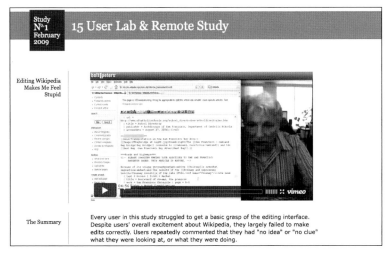

FIGURE 7.7
A sample deliverables page from our Wikipedia study, with project
documents, presentations, and embedded Flash video.

Not All Research Is Research

We want to end this chapter about analysis and reporting—which is about
effectively showing people the value of your research—by commenting on a
bit of conventional wisdom that's come up in the UX and interaction design
communities recently. There's an idea going around that UX research isn't
necessary, because some of the most famous interface design companies
out there—we're looking at you, Apple and 37signals—have famously
claimed that they "don't do research." (An Apple saying goes, "We conduct
user research on one user, and his name is Steve.") This idea intrigues
us because, to be honest, we actually *do* believe that "genius" design and
personal vision should trump user opinions in the design process.

However, we need to make a few things clear: first of all, companies like
Apple *do* conduct research; they just don't *call* it research. Releasing a
product in frequent iterations, as Apple is notorious for, is actually an
audacious form of remote and field research, on a massive scale. The
mountains of reviews, public attention, complaints, crash reports, customer

support issues, and feedback Apple collects (not to mention the Apple retail outlets, which are petri dishes of user behavior) all constitute a wealth of behavioral- and opinion-based data. The same goes for Web sites and services that launch in "beta" for long periods of time: the logs, analytics, and customer feedback are nothing more than open-ended, real-world automated research.

There's absolutely nothing wrong with this approach, but it does presuppose that you'll have a built-in audience to test it on in the first place, and that what you're offering (like a first-generation iPod) is a good enough product in its own right that you'll be able to accomplish your business goals *in spite of* any other usability flaws it might have. Those are very big *Ifs*; you'd basically have to be, say, Apple to pull it off.

Marty Neumeier, author of *The Designful Company*, writes that the key to knowing when to ignore users is to see the difference between *good and weird*, and just plain *weird*. He contrasts Herman Miller's Aeron chair with yogurt shampoo as two products that performed questionably with research, but in which stakeholders chose to move forward despite the known issues. In the case of the Aeron chair, opinions varied about the look of the chair, but everyone said it was comfortable; you could get past how it looked. Yogurt shampoo, on the other hand, was bad on all cylinders; people didn't want *any* part of it.

Finally, just because you might not *need* to do research doesn't mean that you shouldn't or that you wouldn't benefit enormously from it. That's why lots of successful, innovative, and design-minded companies like Google and IBM have relied extensively on both genius *and* UX research to conceive, develop, and improve their products. User research is valuable in designing universe-transforming things only to the extent that you're willing to look for the unexpected, the risky, and occasionally, the sweaty and uncomfortable.

And let's not forget that Apple has made plenty of duds, too. Anyone remember the Newton?

Chapter Summary

- When analyzing session recordings, pay attention to users' computing environment: bookmarks, browser tabs, other open windows, etc.

- Highlight clips should be edited with attention to editing transparency, user privacy, and intelligibility.

- Analysis of task elicitation studies is usually enabled by the tool; common metrics are success rate, time-on-task, and first click.

- Card sort analysis is somewhat subjective and can be time-consuming. Use statistical analysis software tools to speed things along.

- To make your results come across in a presentation, don't just throw everything into a written report—use summaries, highlight clips, screenshots and user images, data charts (for automated studies), sketches and mockups, and simple PowerPoint slides.

- User research isn't the only way to inspire design, and it's not always necessary, but with good, ambitious research goals, it can hardly fail to benefit you.

CHAPTER 8

Remote
Research Tools

W eb-based software comes and goes even faster than regular software, and to be honest, we think 98% of all tools we write about now will either evolve unrecognizably or vanish in five years' time. But we recognize that it's helpful to cover the software tools you'll be using to conduct research, so here we go.

This book isn't an instruction manual, so we'll avoid the obvious "click-here, click-there" tutorial steps. You can find how-to guides on the Web site or in the manual of whichever service or software tool you use. This chapter is more of a reference guide to the best parts of each tool, in our own ludicrously biased opinion. We describe how different tools work for the purposes of remote research, as well as any neat highlights of the tool so that you'll be able to make an informed decision about which ones to use. We cover a handful of the most useful and interesting solutions for screen sharing, screen recording, and automated research, and some tricks we've discovered for getting around their limitations. For most tools, we'll give you a brief summary of their most important features. You can visit their Web sites to learn more if you're curious.

As always, check out http://remoteusability.com for a more up-to-date list of remote research tools and services.

Screen Sharing

You'd think something as simple as screen sharing would be the same for one tool or another, but the tools out there today have plenty of important differences. You not only have to see the participant's screen, but you have to consider these factors:

- Cost

- Security

- Operating system compatibility

- Observing and recording features

- The amount of work the participant needs to do to set it up

- Firewall and spyware blocker compatibility

...and more.

GoToMeeting is one of the most commonly used screen sharing tools simply because it is low cost, works on all platforms, and isn't horrible to use. But other tools might better suit your purposes, so fortunately, you have us to help you sort out this stuff.

Adobe Connect

Adobe makes everything else, so why not screen sharing? Adobe Connect (officially "Adobe Acrobat Connect"), a slightly newer bit of software, is a mid-priced Web-conferencing solution with a few neat twists (see Figure 8.1).

FIGURE 8.1
Adobe Connect: cross-platform, Web-based, quick to set up, and includes integrated video chat.

Neat Features:

- **Integrated video chat.** The most interesting new feature is its integrated webcam chat. If you're running a study in which you know that the users will have webcam access and you'd like to see their faces and desktops simultaneously, Connect is your tool. We're likely to see more and more computers with built-in webcams, so it'll just be a matter of time until you're able to recruit users live and then have the option of seeing their faces minutes later.

- **Choice of monitor or window to share.** Your participants can choose which monitor or window to share. (Still no way to share both, though.)

Things to note:

- The first time you use Acrobat Connect, click Share My Screen. This will prompt you to download and install a plug-in, and that's the only time you'll have to do it. After that, it should always launch using the plug-in rather than a tab in your Web browser.

- There is integrated recording, though again, since it requires you to go through its conference call service, we've always just preferred to record separately.

- There's a crucial difference between Adobe Acrobat Connect and Adobe Acrobat Connect *Pro*, the most notable of which is that the latter costs nearly 10 times as much as the former. Be sure you're ordering the version you want. We made that mistake once, and getting it switched was a real Old Testament struggle.

- It comes with a number of conference-calling options at additional cost.

- To allow your participants to share their screen, you need to select the Auto-promote Participants to Presenters option in the Meeting menu; otherwise, the Share My Screen option won't come up.

- Your participants should enter as Guests. After they enter, you'll need to allow their entry into the room on a little pop-up dialog box that appears in the bottom-right corner of the window, which is easy to overlook.

- Participants have to install the Adobe Connect plug-in at the beginning of the session. It's not large, but it does *not* uninstall automatically at the end of the session. That doesn't mean that you'll be able to see the participants' screen again, but it does mean that you won't be able to say that "nothing gets installed." You may want to walk your participants through the uninstallation process, unless they're okay with keeping it on their computer. We have no idea how to uninstall the thing.

Web site: www.adobe.com/products/acrobatconnect (to buy), connect.acrobat.com/[your connect address] (to join meeting).

Pros: PC, Mac, and Linux compatible; supports simultaneous webcam/ screen sharing.

Cons: Adobe Connect plug-in doesn't automatically uninstall from users' computers.

Requirements: Participants need to have Adobe Flash Player 8 (or later) on their Web browser and a ~450mHz or faster processor. A wired connection is recommended.

Price: For standard version: $39.95/month. For premium version (which includes session recording and other nice things): 32 cents/minute/user on a pay-per-use plan, $375.00/month for 5-user capacity, $750.00/month for 10-user capacity.

Final Word: Solid. The webcam option is pretty interesting but not strictly necessary. For now, we suspect it could be an unnecessary bandwidth and processor hog.

UserVue

TechSmith's UserVue is currently the only screen sharing solution out there that's specifically designed for moderated user research, so it'd be foolish if we didn't mention it here (see Figure 8.2). Unfortunately, it might not be around after mid-2010, but here it is anyway.

Neat features:

- **Observer and moderator video annotating.** While the sessions are ongoing, the observers and moderator are able to type in notes that are integrated into the video file as markers. This capability can be handy for tagging interesting observations and issues for ease of later reference.

- **Observer–moderator and moderator–participant chat.** Instead of throwing moderator, participants, and observers into the same chatroom, UserVue wisely insulates the observers from the participants so that they can chat with the moderator and among themselves, invisible to the participants, preventing disruption.

- **Integrated audio and video recording.** When you finish a session, you get one tidy little recording of the entire session, with the phone conversation synced to the video. How does UserVue do this, you ask, if you're using a normal phone line? Here's what happens: after you enter your phone number and your participant's phone number into UserVue and click Call, your phone will ring. When you answer the phone, UserVue will tell you to dial "1," which will ring the participant's phone as usual. When the participant picks up, the conversation begins just like normal, and both sides of the phone call get routed to UserVue's servers, which syncs the audio with the video.

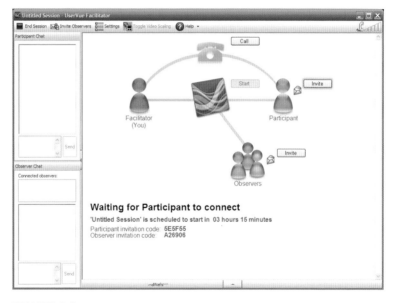

FIGURE 8.2
UserVue: integrated user dialing, video and audio recording, chat, and observing. To be discontinued in 2010.

Things to note:

- TechSmith tells us that it **intends to discontinue UserVue by the second quarter of 2010**, due to lack of demand. That totally sucks, but maybe (if this book spikes interest?) there's a chance TechSmith will keep it around.

- You can't dial outside the Americas directly from UserVue, although there is a way to do it indirectly (see the following sidebar).

- Limit 10 observers at a time.

- UserVue takes anywhere from 5 to 15 minutes after each session to render out the video recording of the session, and you can't start a new session until the rendering is finished. The render time depends on how fast your computer is and how long the session was. You can't batch-render the videos on other computers, as far as we're aware.

- If your or your users' Internet connection is disrupted during the session, or one of you accidently exits the UserVue session, you will see a message in the UserVue window that reads "Your connection to the participant has been lost." Unfortunately, this means that you have to exit the session, render and save the video file, and start with the setup all over again from step 1 to resume the session. Sorry, that's just how UserVue works.

- Like most current screen sharing solutions, UserVue has a decent frame rate and less than a second's lag with a good connection, but not really decent enough to follow along smoothly with any videos that your users are playing. Videos will look choppy on your end.

- There have been rare occasions when the UserVue servers have gone down on testing days, making it necessary for us to resort to a different recording method on short notice. There's really nothing to be done about it, which is why it's important to have backup recording methods on hand.

- The session recordings can be rendered in either WMV format or the proprietary Morae format, which you'll need a TechSmith program like Camtasia to open.

Web site: www.techsmith.com/uservue.asp (to register), http://uservue.techsmith.com (to host/join a session).

Pros: Automatic integrated audio and video recording, observer chat and participant chat.

Cons: No international dialing; PC-only for user, moderator, and observers; 5–15-minute video rendering wait time after each session; have to set up a new session for each user; not cheap.

Requirements: PC for moderator, participant, and all observers; user needs to download and launch an .exe file.

Price: $149/month, 14-day free trial.

Final Word: If you have the money, it's really the best solution for one-on-one moderated testing. Just be sure to have a backup for those cases when it falls short: testing non-PC users, supporting non-PC observers, dialing internationally, and so on.

Testing Outside the Americas with UserVue

Even though you can currently dial only within the Americas in the UserVue interface, here's a sort of roundabout method for using UserVue to test internationally by using Skype:

1.   Launch Skype and begin a UserVue session.

2.   Click Call in the UserVue session window.

3.   Enter your desk phone number in the first field, your Skype call-in number in the second field, and then click Call.

4.   Pick up your desk phone when it rings and then dial "1" when you're prompted to.

5.   You'll get a call in Skype. Answer it.

6.   Mute your desk phone so that the sound doesn't echo.

7.   In Skype, click on Add Caller and enter the participant's number. For international callers, keep the country set to US, type in "011," then the country code, and then the number. You will be talking through your computer and not your phone, so use a USB headset to talk to the participant.

8.   Direct the participant to the UserVue Web site as usual and begin the session.

Warning: Since the audio is being bounced through multiple services (both UserVue and long-distance Skype), the quality can degrade. Test in advance to make sure that the sound is adequate for conducting the session and recording.

GoToMeeting

GoToMeeting is a popular Web conferencing service similar to Microsoft's NetMeeting (see Figure 8.3). It's a little quicker for users to set up; requires only a lightweight, temporary plug-in download for all attendees (no .exe file); comes with free conference calling; and supports observing.

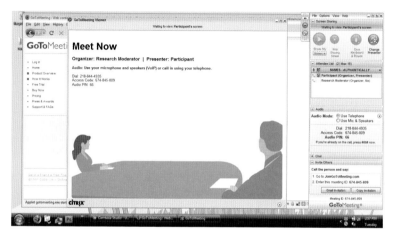

FIGURE 8.3
GoToMeeting: popular Web conferencing with built-in recording, observer and chat support, free conference calling, and remote keyboard and mouse control.

Neat Features:

- **Gives keyboard and mouse control.** This feature not only enables you to view your participants' screen, but also allows your participants to see and even control your screen. This feature enables you to use two remote research techniques we call "reverse screen sharing" and "remote access," which we'll describe in Chapter 9.

- **Free conference calling with TotalAudio.** Although it doesn't boast 100% integrated calling like UserVue does, GoToMeeting does offer a free VoIP conference call system called "Total Audio" to subscribers. The *big* problem with this solution, however, is that it requires users to make a long-distance call to access the conference number. The solution to this is to use a two-line phone to get the users into the call (see the following sidebar).

- **Choice of monitor and window to share.** Your participants can choose which monitor on their computer to share, in case they're using multiple monitors. (Can't share both, unfortunately.) They can also share just a specific window, if there's a privacy issue around sharing the whole window.

- **Pause screen sharing.** You can put the screen sharing "on hold" for whatever reason, without having to disable and reconfigure it entirely.

- **High compatibility.** We've had a lot of success using UserVue with most participant computers, regardless of firewall settings or whether they're international. It also allows for observers using a Mac (but not participants, sadly).

Use a Two-Line Phone to Conference Call with Users in GoToMeeting

Like we said, you probably don't want to require your users to call a long-distance number, so you'll need to use a different way to get around this restriction in GoToMeeting. Use a two-line desk phone and dial into the conference call number; then have your observers do likewise. When you and your observers are ready to begin the session, switch to your phone's second line and dial the user's number and then use the desk phone's conference function (which any decent two-line phone should have) to merge the two calls so that the observers can hear the session.

Don't have access to a dial-in conference call number? You can also have your observers gather in the same room with a speakerphone. Going this route is a little more inconvenient but not too bad if there's only a handful of observers (and if there's only one observer, conference calling is unnecessary).

Important: The observers' phones **must be muted**; otherwise, the participant will hear them!

Things to note:

- Unlike in UserVue, in GoToMeeting the calling is performed separately from the screen sharing, so you can use it to test internationally.

- Also unlike in UserVue, in GoToMeeting you don't need to start a new session for each new user; you can just keep the same meeting window open and let it sit between sessions.

- Although there is chat functionality in GoToMeeting, neither you nor the observers should use it, since your user (who is just another meeting attendee) will be able to read everything, which is distracting to say the least. Use an IM client instead to chat with observers.

- Observers also need to mute their phones so that they won't be overheard by the user during the session. If you're using Total Audio, you can manually mute them yourself if you need to.

Web site: www.gotomeeting.com (to register/host a meeting), http://joingotomeeting.com (to join a meeting).

Pros: Free audio conferencing, recording, ability to give control of keyboard and mouse.

Cons: Observer chat is visible to the participants; the audio conferencing requires users to dial long distance.

Requirements: Users must have a Java-enabled browser; Firefox or Internet Explorer 6+ should work fine. Fully PC compatible; Mac compatible for moderator and observers only (participant must use PC).

Price: $50/month or $468/year, with free 30-day trial.

Final Word: Lacks the nice UX-specific features of UserVue but is more than suitable for remote research, and at one-third the cost of UserVue, we ain't complaining.

LiveLook

Here's a departure from the rest of the screen sharing pack: the online service LiveLook is a 100% browser-based screen sharing solution (see Figure 8.4). There's no software and no need for anyone (participant, moderator, observers) to download or run anything, which means that it can get around most spyware or firewall blocks. On top of that, it can run on any computer with a JavaScript-enabled Web browser, which is basically all of them (Mac, PC, or Linux). It's also pretty good for testing on a slow connection, requiring relatively low bandwidth.

FIGURE 8.4
LiveLook: 100%
browser-based,
cross-platform, no
downloads, quick to
set up.

NOTE HAVE BACKUP SCREEN SHARING TOOLS

Having more than one screen sharing solution on hand is ideal. You'll want to have options in case one fails or doesn't work with the participant's setup, especially if recruits are scarce. LiveLook is a great fallback because of its platform compatibility and ease of setup, but the lack of security and the necessity of giving out account info to strangers might be a deal breaker for you. Either make sure to have backups, or double-check in advance that your users will be able to use your screen sharing solution. Also, since no tool is ideal for every situation, have a few tools on hand that will cover all your bases. We have UserVue, LiveLook, GoToMeeting, and Adobe Connect all at the ready.

Neat Features:

- **Completely browser based.** Like we said, the fact that LiveLook requires no installation or plug-in download makes it the screen sharing option most likely to work on any computer. You can get around all but the most heavy-duty firewalls and spyware blockers.

- **Really quick to set up.** All your user does is visit a link and give you a number; setup takes barely a minute.

Things to note:

- To allow your users to share their screen, you'll have to use LiveLook's "LiveLook Assistant" product. This tool allows you to give a link to your users, which your users click to obtain a session number and begin a screen sharing session. The users give you the session number, allowing you to join the session and view their screen.

- Like in UserVue, in LiveLook you need to begin a unique session for every participant.

- The LiveLook Web site claims that the screen sharing is secure, but we're not really sure how that works, so *caveat emptor*.

- Unlike Adobe Connect and GoToMeeting, LiveLook doesn't resize the participant's screen resolution to fit on your screen; instead, the browser adds scroll bars so that you can scroll around to view different parts of the screen. This limitation can be bad if you need to see the entire screen the whole time, and your participant's resolution is larger than yours.

- Since there's no built-in sound, you'll need to take measures to enable observers to hear the sessions. Use the "Two-Line Phone" technique described earlier in the GoToMeeting sidebar.

- Right now only annual subscriptions are offered for the LiveLook Assistant product. The standard LiveLook product offers per-minute and monthly subscriptions as well, but it requires you to give your users your LiveLook login information to be able to share their screen. If you don't mind giving out that info, the per-minute and monthly accounts are an option. (The sales rep we spoke to mentioned that the company is considering other pricing options but has "nothing solid" yet.)

Web site: www.livelook.com

Pros: The only 100% browser-based, no-download solution we're aware of. Compatible with most operating systems. Very quick to set up. Supports observers.

Cons: No integrated sound, recording, observing—just straight-up, one-on-one screen sharing.

Requirements: Users must have a Java-enabled browser; Firefox, Safari, or Internet Explorer 6+ should work fine.

Price: $360/year for a single account license; $900/year for a "concurrent user" license, which lets you have multiple (6–7) people using the account at the same time.

Final Word: Cheap and bare bones, with a high level of accessibility and compatibility.

Other Screen Sharing Services

Here's a list of screen sharing tools and services that we're less familiar with, but we're sure they're just as lovely as the ones we've listed in the preceding sections.

Cisco WebEx. Similar to GoToMeeting, it's a Web conferencing/online meeting tool that can be used for remote screen sharing. Compatible with Windows, Mac, Linux, Solaris (?), HP-UX (??) and AIX (??!?!). $69/month, $708/year. www.webex.com

Glance. Similar to LiveLook, a browser-based screen sharing solution. Allows for remote access, like GoToMeeting. Up to 100 people in a session at once. Mac and PC compatible. $9.95/day, $49.95/month, $249.00/month (corporate). www.glance.net

Skype. Skype users can view each other's screens in addition to video chat. Requires both parties to have Skype installed and registered. Screen sharing feature is currently Mac only. www.skype.com

iChat for OS X Leopard. On Mac OS X 10.5 and up, there's a feature in iChat that allows two iChat participants to video chat, share screens, and give remote desktop access. Comes standard with Mac OS X 10.5 (Leopard) and above. www.apple.com/macosx/what-is-macosx/ichat.html

QualVu. This is a brand new one, which we've never used before. Uses users' webcams to run "online video qualitative research" studies, in which you can video chat with panel users, watch them interact with and respond to your interface, and take notes and instantly create highlight clips. Sounds weird and interesting. www.qualvu.com

Yugma. Web conferencing service with lots of nice Pro features: real-time whiteboard and annotation collaboration tools, building in recording, free teleconferencing, file sharing. Integrates with Skype. Installation required for both hosting and joining a session. Basic screen sharing with up to 20 participants is free; Pro is $14.95 / month. www.yugma.com

Yuuguu. Screen sharing and collaboration software. Can import IM contacts from other services (AIM, GChat, Skype) Requires download and install; Mac, Linux, and PC compatible. www.yuuguu.com

NOTE YOU CAN'T SHARE MULTIPLE MONITORS YET

As far as we're aware, no screen sharing tool out there allows you to see more than one of the participants' monitors at once, except for VNC, which is probably not an ideal remote research tool, since it takes a long time to set up and is pretty invasive to install on participants' computers. A few tools, such as GoToMeeting and Adobe Connect, allow your participants to select which monitor they'll share.

Recording

Recording your sessions is a comparatively simple and straightforward consideration. If you're using a screen sharing tool that doesn't provide recording, you'll need two things: a way to get the audio and video of your session both going through your computer and a program that captures the audio and video output of your computer. The recording software often takes a bit of processing power, so make sure your computer is swanky enough to run it alongside whatever else you have running.

Camtasia Studio/Camtasia for Mac

Our recording solution of choice, Camtasia Studio (by TechSmith, makers of UserVue), is a simple tool that can record the video and audio output on your computer in just a few clicks (see Figure 8.5). We like it because it gives you the option to record just a portion of your screen or just a single window on your computer, which helps keep the recording file size trim.

FIGURE 8.5
Camtasia Studio: record computer video and audio, full screen, or portions of the screen.

Things to note:

- The trickiest thing about using Camtasia is making sure the audio is being recorded. If you're using a voice chat service like Skype to chat with your users through your computer, the audio should automatically be picked up by Camtasia. Otherwise, you'll need to route the signal from your phone to your computer using a phone tap and amplifier. Once you've got it hooked up, go to the Tools menu, select Options, click the Audio tab, and you'll be able to test if the telephone sound is going through or not. If it's not, you probably need to change the audio input device setting, which is also under the Audio tab.

- Be sure to set the video files to save as AVI, which is compatible with most media players, as opposed to the default proprietary CAMREC format, which plays on almost nothing. (To do this, go to Tools › Options › Capture tab › Save as .avi.)

- The Mac version has a slightly different set of options, so check the Web site to make sure it has what you need.

- There's also an Automatic file name option (Tools › Options › Capture › File Name Options) that will automatically add a prefix and a number to the filename of each video file that gets produced. We recommend using the study name as the prefix; it provides an easy way to numerically organize the session recordings. So, for example, if the prefix is RemoteStudy, your first recording would come out as "RemoteStudy-1.avi.," the second would be "RemoteStudy-2.avi," and so on.

- Be careful not to obscure the window you're recording with any other windows.

- We've never dealt with this before, but some people have reported problems where after four or five sessions, sometimes Camtasia wouldn't record the next one properly, erroneously citing a lack of hard drive space. Copying the files onto an external hard drive after each session seemed to prevent this problem.

- We've found that there's a bug in older versions of Camtasia (up to at least version 3), which prevents the audio from merging into the recording if the file size grows past a gigabyte, which makes the file effectively useless.

- As always, extensively test the tool with real telephone conversations, making sure both sides of the conversation are being picked up, the volume level of the recording is clear, and the video file works with your media player.

Web site: www.techsmith.com/camtasia.asp

Pros: Reliable, simple.

Cons: Older versions have been buggy.

Requirements: *Windows version:* Microsoft Windows XP or later, Microsoft DirectX 9 or later, 1.0 GHz processor (2.0 GHz recommended), 500MB RAM minimum (2GB recommended), 115MB of hard disk space for installation.

Mac version: Mac OS X v10.5.6 or later, Mac computer with an Intel processor, Quartz Extreme support, 1GB of RAM (2GB recommended), ~4GB of available disk space, QuickTime 7.5.5 or later, CD drive required for installation.

Price: $299 for PC; $149 for Mac.

Final Word: A bit pricey for software, but you'll have to buy it only once, and it'll cover you for screen recording until the very end.

NOTE MAKE SPACE FOR RECORDINGS

Make sure that you have enough room on your hard drive to store the recordings you generate during the sessions. Depending on the recording quality, a 40-minute session recording can be upwards of a gigabyte. Another option, if you're low on space, is to get an external hard drive to transfer over each recording file after each session. And be sure to securely back up those files whenever you get the chance.

CamStudio

Turns out that some wonderful intrepid programmer out there has created a 100% free, open source feature clone of Camtasia called CamStudio (see Figure 8.6). All the core functionality of Camtasia is there: the option to record a portion of the screen on a specific window, the ability to capture audio and video, and so on.

Web site: http://camstudio.org

Pros: Free; adjustable recording quality; able to convert AVI files to more compact SWF flash video.

FIGURE 8.6
CamStudio: a free open
source alternative to
Camtasia Studio.

Cons: PC only; as a free product, there's no professional customer support, but there is a support forum on the Web site.

Requirements: Windows XP or later; system requirements not listed but probably the same as Camtasia.

Price: Zero (but pitch in a few bucks, eh? They earned it.)

Final Word: The price is right. Just make sure it works perfectly for you because you won't have anyone to call if it breaks.

iShowU HD

And now, for you Mac users: shinywhitebox's iShowU HD is a very full-featured recording solution, with tons of nice video and audio recording options, options to record portions of the screen or windows, and customizable hotkeys (see Figure 8.7). A Pro version of the software has even more options, such as a "Low CPU mode" to help avoid slowing down your computer.

Things to note:

- Set the Mouse Mode menu to Fixed. Otherwise, you'll notice that the recording follows your mouse around on the screen.

- Also be sure to disable the Keys switch in the toolbar, or else everything you type during the session will be plastered across the video.

- If you're using a phone tap to route audio to the computer, go to Advanced › Sound. Make sure the "Record sound from input device" box is checked and the menu is set to Built-in Audio. If you're talking through the computer with Skype or some other voice chat program, check "Record audio from applications."

- iShowU HD is set by default to stop the recording automatically when your hard drive has only 200MB remaining. You can change this setting in the preferences.

- As with the other recording solutions, be careful not to obscure the window you're recording with any other windows.

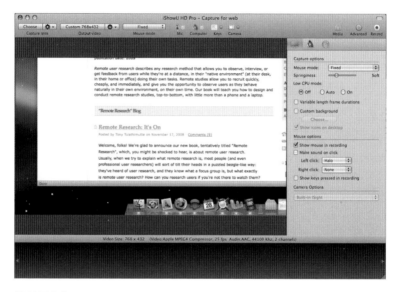

FIGURE 8.7
iShowU HD Pro: lots of cool recording features.

Web site: www.shinywhitebox.com/ishowuhd/main.html

Pros: Feature rich; lots of video and audio rendering options; adjustable video quality; Pro version has Low CPU mode; cheap.

Cons: Tricky default settings; lots of options make it slightly harder to learn.

Requirements: Mac OS X 10.5 Leopard or later. (An earlier version of the product, iShowU Classic, is still available for Tiger 10.4.) System requirements not listed.

Price: $29.95 for standard, $59.95 for Pro, which includes extra features (watermarking, key recording, audio mixing, low CPU mode, Final Cut compatibility).

Final Word: If you're going to record on the Mac, we wouldn't recommend anything else.

Other Recording Options

Adobe Captivate. Expensive, many-featured screen recording tool. PC only. $249. www.adobe.com/products/captivate

Quicktime Pro. Mac OS X's video player also has audio/video recording capabilities, as well as some basic video trimming features. Mac and PC. $29.99. www.apple.com/quicktime/pro

Screenflow. Records screen, webcam, mic, and computer audio simultaneously. Also includes editing and overdubbing tools. Can export to WMV with an additional plug-in ($49). Mac 10.5+ only. $99. www.telestream.net/screen-flow

Snapz Pro X. Records screen and audio to QuickTime format. Mac only. $69. www.ambrosiasw.com/utilities/snapzprox

WMCapture. Basic screen and audio recording. PC only. $39.95. http://wmrecorder.com/wm_capture.php

ZDSoft Screen Recorder. Screen and audio recording, effective at recording anything that uses special rendering, like video playback and games. PC only. $39. www.zdsoft.com/products.html

For even more tools, visit the screen capture page at the All Streaming Media Web site: http://all-streaming-media.com/record-video-from-screen.

> ███ **NOTE** USE AS FEW TOOLS AS POSSIBLE
>
> The more tools you use to test, the more likely one of them is going to mess up, and the harder it will be to figure out what's going wrong. Using Skype for voice chat with a participant, a screen recorder, an audio recorder, an IM client for observer chat, a stopwatch to keep time, an Excel spreadsheet for notes, etc., you're going to be at the outer limits of professional reliability. If possible, try to use a setup that uses as few tools as necessary to get the job done, and always have backup alternatives on hand in case any part of the setup fails.

Automated Tools and Services

When it comes to automated research for a given task (card sorting, surveys, etc.), most tools and services you use will be relatively similar, with only a few features available or missing. We already covered the methods for using automated services in Chapter 6, so in this section we'll just provide you with a wealth of services to check out, with features, pricing info, and links to relevant Web sites. We've divided them up here into the same categories as in Chapter 6.

Note: These are not exhaustive feature lists; they're summary paragraphs emphasizing noteworthy features of each tool. And we probably missed a few tools; we'll be adding more to http://remoteusability.com as we discover them.

Task Elicitation and Analytics Tools

Click Density. A UK-based service offering interactive heatmaps, which you can segment and filter along several variables (like user path, browser, screen size, etc.) and export into presentations. Requires a line of code to be inserted into your Web site. Free basic account, $5/month (10,000 clicks, 1 site), $20 (45,000 clicks, 10 sites), $100 (250,000 clicks, 25 sites), $200 (1,000,000 clicks, 50 sites), $400 (5,000,000 clicks, unlimited sites). www.clickdensity.com

ClickHeat by LabsMedia. This is a humble open source heatmap program, and as such, it's pretty bare bones on features—just a heatmap overlay that goes over an image of your Web site. It also requires much more tech prowess than the commercial solutions. Still, if free is what you want, this is it. www.labsmedia.com/clickheat/index.html

ClickTale. Records user interaction with forms, links, and other elements on your Web site (see Figure 8.8). Generates heatmaps of places users are clicking and scrolling, how long users spend, and how many page views each section gets. This tool can also record real-time movies of user behavior for later playback. There's no task elicitation involved; it just tracks natural user behavior. Requires a few lines of code to be inserted into your Web site. Free basic account, $99/month (email support, 10 domains, 20,000 page views), $290 (priority support, 25 domains, 80,000 page views), $790 (phone support, unlimited domains, 240,000 page views). www.clicktale.com

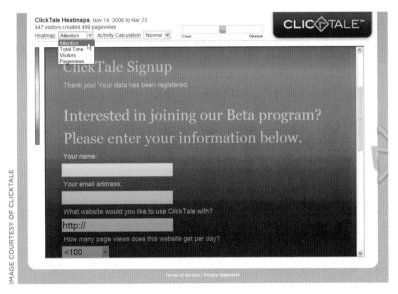

IMAGE COURTESY OF CLICKTALE

FIGURE 8.8
ClickTale's "Attention" heatmap, which illustrates where users click and how far they scroll down the page.

Clixpy. Tracks mouse movement, scrolling, and form input, capturing it in video format. Requires pasting a few lines of JavaScript into your site code to install. First 10 video captures free; 100 for $5, 200 for $10, 600 for $20, 1,000 for $30. www.clixpy.com

CrazyEgg. CrazyEgg has a bunch of tools for visualizing the clicks you get on different elements of your site, including heatmaps, colorful clickmaps, and site overlays. You can share the results with others using a custom link, export the results to Excel, and receive notifications about the status of your study via email and RSS. Requires one line of code to be inserted into your Web site; $9/month (10,000 reports, 10 pages), $19/month (25,000 reports, 20 pages), $49/month (100,000 reports, 50 pages), $99/month (250,000 reports, 100 pages). www.crazyegg.com

Chalkmark by Optimal Workshop. Allows users to complete tasks on static images, providing heatmap feedback for prototypes and mockups. Free account provides surveys of up to 3 tasks; $109/month, $559/year. www.optimalworkshop.com/chalkmark.htm

Google Analytics. Most people don't think of this service as a "research tool," but that's exactly what it is. The only difference is that none of the tasks are elicited. It's a great tool for slicing and dicing your Web traffic data, and it has functionality that allows you to also track which paths users are taking through your site, which is great for monitoring "funnel tasks" (linear, step-by-step tasks, going through a checkout process, registering for an account, etc). Best of all, it's free. www.google.com/analytics

LEOTrace by SirValUse. Analytics service and tool providing log data, mouse movements and clicks, and data reports such as user behavior video and heatmaps. SirValUse offers full-service consulting as well as licensing its tool. Price upon request. www.leotrace.com

Loop11. Pretty straightforward task elicitation: a browser bar prompts users to perform a task on any interface that can be displayed in HTML, and the users indicate whenever they have either completed the task or have abandoned it (i.e., given up). All interactions captured into real-time reports. Automated analysis includes task completion rate, time on task, first click, clickstream, most frequent success/fail points, etc. Available in 40 languages, so international testing is a go. It's currently only in private beta, which means you can register by invite only. www.loop11.com

m-pathy. A remote usability tool that tracks mouse movements and clicks without installing anything on the user's computer. It's in German, so we don't actually know what the heck the Web site is saying, but from the looks of it, it appears to offer heatmaps of clicks and mouse movement and also mouse tracking. http://m-pathy.com

OpenHallway. Prompts users with instructions and tasks and then records their behavior on the site and spoken think-aloud comments into a sharable streaming video recording. Sessions are limited to 10 minutes. No software to install for participants or researchers. Firefox, IE 7+ and Safari supported; participants must have Java 1.5+ enabled. Recruit your own participants by sending them a link to participate. Free 30-day trial; $49/month for 3 hours of recording storage, $99/month for 10 hours, $199/month for 30 hours and the ability to download videos. www.openhallway.com

RelevantView. RelevantView is a user research consulting company that also offers its automated research tools for self-use. Its task elicitation tools allow you to track time on task, task completion, and click paths. These tools also allow you to do competitive research and test prototype sites. Pricing upon request. www.relevantview.com

SMT (Simple Mouse Tracking) by Luis Leiva. An open source project, providing mouse movement and click tracking functionality; it also tracks scroll bar, mouse wheel, and highlighting activity. Like ClickTale, it allows you to replay user behaviors in real-time, and the activities can also be exported to XML. If you're tech-savvy, you can customize it. It's free, licensed under Creative Commons. http://smt.speedzinemedia.com/smt

Testled by '80s Business Guys. Record user behavior with no installs or Javascript required. Participants can use any browser or OS. Currently in invite-only beta. http://testled.com

Usabilla by Usabilla BV. Basic task elicitation on Web sites or static images; you have to provide your own recruits. Data are visualized as "datapoints, heatmaps, cluster clouds, notes." Also includes screen annotation. Currently in free beta. www.usabilla.com

Usability Exchange by KnowFaster.com. Automated task elicitation for testing disabled users (blind, dyslexic, etc.); disabled participant pool provided by Usability Exchange. 5 users and 4 tasks = £299, 10 users and 4 tasks = £599+, custom number of users and tasks = at least £299, simple questionnaire = £199 www.usabilityexchange.com

UserTesting.com. Task elicitation in which users' spoken think-aloud comments are captured and synced with videos of their Web site behavior as they perform tasks of your choosing. Videos are downloadable, editable, embeddable, and can be annotated with comments. You can also see users' demographics and system info. Users are recruited by UserTesting.com from a preselected panel; $29/user. www.usertesting.com

UZ Self-Serve Edition by UserZoom. UserZoom, an automated research company, has its remote research tools available for do-it-yourself studies. This Web-based tool allows you to manage multiple automated research projects; gather clickstream data; prompt users to perform Web site tasks, card sorts, and surveys; and recruit users from a panel, your email contacts, or your own Web site. It also offers full-service UX testing. The tools are currently available only by yearly subscriptions; pricing available on request. www.userzoom.com

WebEffective by Keynote. Another Web-based tool for conducting in-depth customer experience, branding, and market research studies. Users answer survey questions and complete tasks in pop-up windows, with no download required. Keynote employs a panel to provide quantitative clickstream and behavioral data, survey feedback, and task completion data. You can also intercept users from your own Web site. Studies are available in 17 languages. Pricing available on request. www.keynote.com

Webnographer by FeraLabs. Full-service automated research company that will soon be releasing self-service tools. Tracks clickstream data, allows you to present questionnaires to the user. No code needs to be modified on the site being tested, and no user download necessary. All you do is send participants a link. Cross-platform and cross-browser compatible. Full-service research is $1,200/day; average time is about 1.5 days but can range from 1–5 days, depending on the complexity of the interface and the number of participants. Can recruit from your Web site, Webnographer panel, or your email contacts. Pricing to be announced. www.webnographer.com

Survey Tools

We repeat: Surveys aren't really within the domain of remote UX research; mostly, they're a tool you use to gather opinions, not record behavior. But since so many of us are tempted to put up surveys, we're including some of the best tools here in the hope that you'll use them only sparingly, to accompany your behavioral research.

HaveASec. This is one of the first entries in a sure-to-be robust ecosystem of mobile survey research tools, targeted at iPhone developers gathering feedback for their apps, but it's compatible with some other devices as well. Developers can either embed the survey within their iPhone app or provide a link somewhere within the app that takes users to a mobile-optimized Web site containing the survey. We see great promise in using apps like these to live-recruit participants for mobile remote research studies, which we discuss briefly in the next chapter. Free basic account, $50/month for premium (up to 1,000 responses per survey), $500/year for platinum (up to 5,000 responses per survey). www.haveasec.com

PollDaddy. A nice, cleanly designed survey tool with 11 question types, conditional branching, full foreign language support, multipage surveys, and the ability to customize surveys with HTML. Charges by the response, rather than on a subscription basis. Free basic account, $200/year (1,000 responses), $899/year (10,000 responses). www.polldaddy.com

RelevantView. In addition to task elicitation, you're able to design your own surveys with branching logic and view and export results in summary, raw, and cross-tabulated formats. Surveys can be designed in all non-Asian languages. Pricing upon request. www.relevantview.com

SurveyGizmo. Skip logic and survey branching, browser redirects, a handful of question types (see Figure 8.9). Tools to filter, cross-tabulate responses, and remove duplicate entries. You can export data to text and Excel and export reports to Excel, Word, and PDF. Reports can be visualized in three kinds of charts. The survey code comes in JavaScript, HTML, and iFrame, so it can be embedded into both Web sites and email. Twenty types of questions. You can manage and track email invitations to participants, as well as send out reminders and follow-up emails. Template surveys available for beginners. Like PollDaddy, it charges by the number of responses, and it also offers discounts to nonprofit companies. Free basic account and 14-day trial; $19/month (1,000 responses, 5,000 invites), $49/month (5,000 responses, 10,000 invites), $159/month (50,000 responses, 10,000 invites), $599 /month (1,000,000 responses, 100,000 invites). www.surveygizmo.com

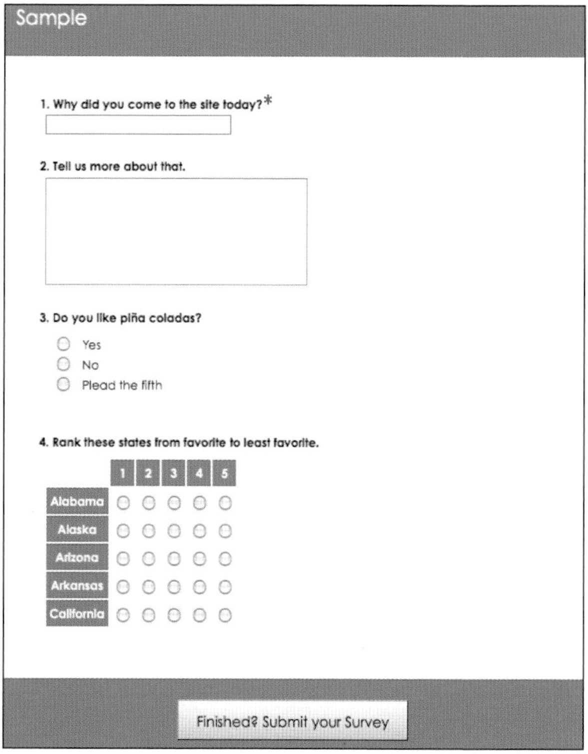

FIGURE 8.9
Sample SurveyGizmo survey. Remember, surveys give opinion-based feedback, not behavioral feedback!

SurveyMonkey. One of the most popular online survey tools out there. When designing a survey, you can choose from 20 kinds of form questions, support for all languages, response error checking (requiring users to input valid answers). This tool has a built-in manager for sending out and keeping track of survey invitations, as well as a pop-up feature to get people to fill out your survey. You can filter and cross-tabulate the results of the survey, share them with others by sending them a link, and export them to Excel, XML, and HTML format. Free basic account, $19.95/month (monthly pro), $200.00/year (annual pro). www.surveymonkey.com

UZ Self-Serve Edition. Takes an interesting approach to surveys: rather than displaying a pop-up when users arrive at the Web site, it makes a browser window that users see after closing down the site, so the survey is more like an exit survey. This allows you to see whether users are able to complete whatever tasks they were on the site to perform, as well as assess user satisfaction. Pricing upon request. www.userzoom.com

WebEffective. Also offers basic survey tools, in addition to the task elicitation tools listed previously. www.keynote.com

Zoomerang. A survey tool with all the usual nice features: skip/branching logic, exporting, and the ability to post the survey to your site. It also maintains a panel of survey respondents if you don't feel like recruiting from your own Web site. With the premium account, you're able to create mobile surveys that users can respond to on their mobile devices. Free basic account, $199/year (pro), $599 (premium). www.zoomerang.com

For even more commentary on what to look for in online survey tools, check out Idealware's "A Few Good Online Survey Tools" article: www.idealware.org/articles/fgt_online_surveys.php

Card Sorting Tools

OptimalSort by Optimal Usability. One of the first and foremost online card sorting services, OptimalSort uses a simple drag-and-drop interface to allow participants to sort cards into groups and also make additional comments in a text field at the bottom of the screen (see Figure 8.10).

You're able to customize the study introduction, a preliminary questionnaire, the study instructions, the cards and categories, a debriefing "Thank You" message, and you can ask for users' contact info for incentive payment. Helpful study templates are available for beginners. The results are loaded into an exportable table, which you can organize by card and category labels, participants, and comments. You can also view the individual users' raw data. The nice folks at OptimalSort offer free basic advice and support for using their tool (provided it "doesn't take a lot of their time"), and also paid consulting services. Free basic account, $109 (30-day subscription), $199 (90 days), $559 (annual). www.optimalsort.com

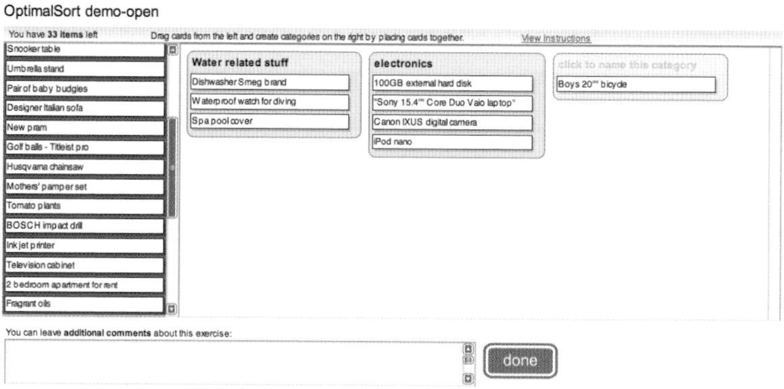

FIGURE 8.10
Optimal Sort: a simple drag-and-drop online card sorting tool, with templates and exportable data (which can be used with the Card Sort Cluster Analysis Tool).

RelevantView. Allows users to drag and drop pictures, text, symbols, diagrams, and images. The Web site seems to indicate that it supports only the closed-sort method. Pricing upon request. www.relevantview.com

WebSort. Like OptimalSort, WebSort gives users a simple drag-and-drop interface to sort cards, create groups, and comment. It allows you to customize the instructions and thank-you message, share the results with others, import studies from OptimalSort, and view the results of the study in tree diagram form in addition to table form, with the ability to export them to an Excel file. Free limited account, $79 (single study), $299 (5 studies), $199/month (unlimited studies). http://websort.net

UZ Self-Serve Edition. Also packs a card sorting tool, allowing users to sort up to 100 cards into up to 12 categories. You can ask context questions to find out why users have decided to categorize things the way they do, and you can also include other UserZoom automated testing elements besides the card sort in the same study. The data can be viewed either raw, in a matrix, or in a dendrogram. Pricing upon request. www.userzoom.com

Other Automated Research Tools

Card Sort Cluster Analysis Tool 1.0. An online tool you can use to take exported results from OptimalSort and organize them instantly into an affinity chart. Handy! Free! www.userpoint.fi/card_sort_cluster_ analysis_tool

Ethnio. Sorry, we couldn't resist hyping our own product again. We think it's worth another mention here, not only because it's the only tool out there specifically made for live recruiting, but also because it has a hidden benefit: it allows you to make an extract of every visitor who's filled out the recruiting screener, producing an interesting bit of quantitative data, even if you're just recruiting for small-sample moderated research. Since Ethnio exports the data into a spreadsheet, you can organize that data later by response. Note that since respondents are self-selected from your Web site (rather than randomly sampled), these data aren't up to the standards of rigorous statistical analysis; still, they're interesting bits of data to include along with your qualitative findings. Requires a line of JavaScript to be inserted into your Web site. First 20 recruits free; $400/200 recruits, $2,000/2,000 recruits. http://ethnio.com

Google Docs (Forms). Allows you to make very simple forms for surveys and other online feedback. Building and editing forms are a breeze, and you can send a direct link to the forms or embed them into your site with a bit of code. The results are loaded automatically into a Google spreadsheet, nicely formatted and timestamped. The reason we wouldn't recommend this tool for recruiting or sensitive data gathering is that Google technically possesses all the data, and we haven't looked into how secure or private those data are. Maybe they're fine! At any rate, this tool is 100% free with a free Google account, and it seems the obvious choice for gathering nonpersonal feedback automatically. http://docs.google.com

iMarkIt by ITracks. A concept testing tool for screen annotation. Participants can mark up video, Web sites, images, and text with emoticons, arrows, colors, and text boxes. Seems as though participants are recruited from a panel, but we couldn't tell. Pricing upon request. www.itracks.com/Products/ConceptTestingiMarkIt.aspx

TreeJack (Beta) by Optimal Workshop. This is sort of but not really like a task elicitation tool. Instead of having users perform tasks on the actual Web site, you load your Web site's navigation structure into TreeJack and have users perform tasks on a hierarchical tree diagram that represents your site structure. For example, you can ask users where in the navigation they would find "information about usability research," and they navigate the site's structure, indicating which page they would expect to find it on. This is a handy tool for information architects who want to test the intuitiveness of their navigation structure. Currently in beta, so it's free. www.optimalworkshop.com/treejack.htm

Wufoo. We covered this all-purpose automatic form-building tool in Chapter 3, showing you how you could use it to recruit and obtain consent, but it can also be used to develop surveys that can be embedded right into your Web site. You can search through the entries (i.e., user responses) and export them to Excel and text formats. If you're using this tool to recruit from your Web site anyway, it seems like a logical choice for basic surveys. Free basic account, $9.95/month (10 forms, 500 entries, 200MB storage), $24.95/month (unlimited forms, 3,000 entries, 500MB storage), $69.95/month (unlimited form, 15,000 entries, 1GB storage), $199.95/month (unlimited forms, 100,000 entries, 3GB storage). http://wufoo.com

Chapter Summary

- The most important factors to consider in a screen sharing tool are cost, security, OS compatibility, observing and recording features, ease of setup, and firewall/spyware compatibility.

- Your recording solution shouldn't slow down your computer, and it should output to a standard video format and record audio and video. Remember to have enough hard drive space to store recordings.

- With automated tools, pay attention to pricing and cost, setup requirements (for both you and the user), what data can be captured, qualitative feedback options, and results formatting.

CHAPTER 9

New Approaches to User Research

N ow that we've run through the standard varieties of remote research, we'd like to show you a few variations of remote research we've come up with in the past, which adapt remote techniques to address special testing conditions or to improve on existing research methods. These variations demonstrate how adaptable remote methods are, and we hope you'll be encouraged to make your own modifications and adjustments to the basic methodological approach we've laid out in this book.

Each section in this chapter features a mini-case study of the research projects in which we developed each technique, so you will not only hear about the technique, but also get some insight into the way we devise new remote research techniques.

Top Secret (Reverse Screen Sharing and Remote Access)

In some cases, instead of seeing your users' computer screens, you need to have them see or take limited control over one of yours. This type of control is useful when you need to show your interface to your users but either can't or don't want to give them direct access to it on their own computers. Let's say you want users to try your prototype software but don't want them to be able to install it on their computers. When you use reverse screen sharing and give them remote access to one of your computers, they can use the interface without having possession of the interface code or files in any way.

In 2008 we designed an international remote study for a multinational bank. The bank wanted to keep a new Web interface relatively secret until its release and didn't want its fairly advanced prototype to be floating out in the Internet wilderness. Since the target audiences were scattered across the Pacific Rim (the Philippines, Australia, Hong Kong) and our clients were based in different countries as well, remote testing was definitely the most attractive testing solution.

We needed to come up with a way to keep the Web prototype safe on our computers while still allowing users to interact with it normally. Ordinarily, you might be inclined to make a Web-based mockup that the users could access temporarily via password protection, but if you're on a tight deadline, that approach won't always be an option. We needed to test the interface ASAP, so we opted for reverse screen sharing.

Method

First, we loaded the prototype onto a second Internet-connected computer (which we'll refer to as the "prototype computer") and connected it to a monitor of its own, next to the moderator's computer setup. We used GoToMeeting for screen sharing and hosted the screen sharing session on the prototype computer. The moderator contacted the participants and instructed them to join the GoToMeeting session as usual. Once participants had joined, the moderator used the "Give Keyboard and Mouse Control" function on the prototype computer, specifying that the participants could access only the window with the prototype in it (see Figure 9.1). After that was all set up, participants were able to view and interact with the prototype on their own computer screen, though they were really just remotely viewing and controlling the prototype computer's desktop (see Figure 9.2).

FIGURE 9.1
Giving mouse and keyboard control to the participant in GoToMeeting.

FIGURE 9.2
This Mac user is taking control of a Windows PC remotely.

Another perk of remote access is that if your interface requires any sophisticated software to run properly, you can set it up in advance without requiring users to download, set up, and install it themselves.

On the other hand, this method does takes slightly longer to set up, and depending on the strength of the Internet connection, there may be some slight (~1 second) lag, causing the interface to seem sluggish. The sluggishness is more of an issue when testing internationally. Run a pilot session in advance to make sure that the approach is viable or arrange in advance for your users to be on a fast and reliable Internet connection. (We arranged in advance for our Hong Kong users to be at a location with a solid connection.) This method also slightly restricts the types of interfaces you're able to test effectively. If you're testing a video game, for instance, the lag might impact the experience too much to provide a decent user experience.

Other than those limitations, remote access is a great way to keep your interface airtight, without too much extra hassle or setup. Really, all it takes is an extra computer to serve as the prototype computer.

Final note: any screen sharing solution that allows you to give users remote access to your computer desktop can be used for this method, but be sure that you use a method that lets you easily and permanently disable the remote access once the session has ended.

Mobile Device Research (Smartphones)

It doesn't take a genius to see that mobile computing devices are taking over. Our relationship to technology is ever moving toward the academic portents of ubiquitous computing; at the very least, we're going to have some super fancy cell phones. The richest and most exciting future for remote research is on mobile devices, but currently the technology to conduct the research isn't completely there yet. There's been lots of academic and professional work on how to properly research mobile interfaces, but most of these studies have involved cumbersome apparatuses, structured in-lab usage, expensive ethnographic interviews, or quantitative data gathering.

One study we find very promising isn't a user research study at all, but a study on happiness. Harvard researcher Matt Killingsworth and Visnu Pitiyanuvath's "Track Your Happiness" study uses iPhones to gather random experience samples from people. Participants are notified by text message over the course of the day; the message directs them to a mobile-friendly Web site that asks them questions about what they were doing and how they felt in the very moment before they received the notification. After each survey, participants are displayed some preliminary results—for example, how their happiness has correlated with whether they were doing something they wanted to do (see Figure 9.3). This is a fantastic example of time-aware research because it collects data that's dependent on the native timeline of participants.

This research approach—having users self-report their circumstances and feelings at random intervals during the day—is known as "experience sampling," and what makes it so promising is that it plays to the strengths of remote research. By getting on the participants' time, researchers are able to get a more real sense of what the participants are *really* thinking and feeling in their real lives, rather than in a controlled lab environment.

And by simply adapting currently existing technology (SMS, smartphone browsers) to suit research purposes, it doesn't force participants to work around awkward and unrepresentative testing equipment.

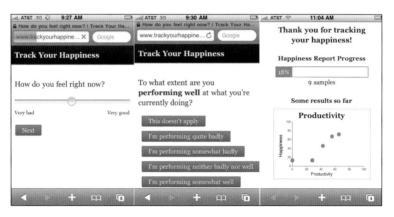

FIGURE 9.3
The "Track Your Happiness" iPhone survey: screenshots of questions and preliminary results.

We're beginning to see some services and Web applications that will make this type of research feasible; in the preceding chapter we introduced the HaveASec app, which allows iPhone developers to push surveys to their mobile app users and receive responses in real time. This capability could easily be applied to mobile live recruiting. You could call up users on their mobile device right when they're in the middle of using your application, so you can learn about not only what they're doing with it, but where they are, how their reception and location are affecting their usage, and all those other excellent native environment insights. Mobile screen sharing is also beginning to make inroads. PocketMeeting, Veency, and Screensplitr are emerging examples of mobile screen and interface sharing.

Of course, there will be many challenges for mobile remote research, not least of which is convincing mobile users who are likely busy, on the go, or in a noisy public space to participate in a live study. You'd also need to find a way for the users to both talk on their phone and use the phone app at the same time (a headset would work). Safety would naturally be a consideration, as would battery life, reception, and social factors. The participants may be in a potentially awkward social space (library, dentist chair, restroom).

But as with remote research over the Web, UX research methods for mobile interfaces will improve and become easier as the technology advances. Improved phone video cameras, Web video streaming and video chat, VoIP calling, automated screen-recording and activity monitoring, GPS location tracking, and third-party apps are all likely developments that will prove to be invaluable to understanding how real people use their mobile devices. And accessibility is a factor, too. Right now, only a tiny fraction of the population owns phones capable of this type of research, but as smartphones become cheaper and more practical, we can begin to get a broader picture of how people live.

One-to-Many Remote Moderating (Video Games)

If you recall from Chapter 1, we suggested that lab testing can be preferable over remote testing when you're unable to give remote access to your interface for any reason. We designed and conducted a series of such studies in 2007 and 2008 for Electronic Arts' highly anticipated PC game, *Spore*. There were a number of reasons why the game couldn't be tested remotely, and chief among them was that the game was still in the development process and couldn't be released to the general public for confidentiality's sake.

The most practical alternative was to conduct the testing in our lab, but we decided that it was important to maintain the native environment aspect as much as possible. A video game is a special kind of interface in which preserving a comfortable and relaxed atmosphere is crucial for producing a realistic user experience. We attempted to replicate aspects of a native environment, minimizing the distraction by other players, research moderators, or by arbitrary tasks not related to gameplay. The result was what we call a *simulated native environment*, or a lab setting that resembles a natural usage setting as much as possible (see Figure 9.4).

FIGURE 9.4
Participants each had their own isolated stations, tricked out with a Dell XPS laptop, mouse, and microphone headset through which they communicated with the moderators.

Method

For each session, we had six agency-recruited participants come to our loft office, where they each sat at their own desk equipped with a laptop, a microphone headset, and a webcam. We instructed them to play the game as if they were at home, with the only difference being that they were to think-aloud and say whatever was going through their minds as they played.

Elsewhere, our clients from Electronic Arts (EA) and our moderators were stationed in an observation room, where we projected the players' game screens, the webcam video, and the survey feedback onto the wall, allowing us to see the users' facial expressions and their in-game behaviors side by side (see Figure 9.5). Using the microphone headsets and the free game chat app TeamSpeak, we were able to speak with players one on one, occasionally asking them what they were trying to do or prompting them to go more in depth about something they'd said or done in the game. This method allowed us to conduct six one-on-one studies at the same time, which we called "one-to-many moderating."

FIGURE 9.5
In the observation room, the moderators sat side-by-side with the observers, monitoring all six players' expressions, gameplay behaviors, and survey answers at once.

EA was interested in collecting some quantitative data as well. When participants reached certain milestones in the game, they would fill out a touchscreen survey at their side, answering a few questions about their impressions of the game. These surveys were designed by Immersyve, a game research company in Florida, basing their quantitative assessment of fun and enjoyment on a set of validated metrics called Player Experience of Need Satisfaction (PENS).

Equipment Setup

A great deal of the work that went into preparing for this study went into figuring out our incredibly baroque hardware and equipment setup. Each of the six stations needed to have webcam, gameplay, survey, and TeamSpeak media feeds broadcasted live to the observation room; that's 18 video feeds and 6 audio feeds. Not only did the two (yes, two) moderators have to be able to hear the participants' comments, but so did the dozen or so EA members. On top of that, all the streams had to be recorded for later analysis, which required hooking up digital recording units to each station (software recording solutions performed poorly), enabling TeamSpeak recording at each station, and using the webcam software's built-in recording function.

Your setup doesn't have to be quite this complicated. Our multiuser approach was only necessary due to the volume of users we had to test, but if possible, take it easy and test one or two users at a time, with a single moderator. Here's a step-by-step guide on how to set up a simulated native environment lab for a single user.

First, set up the observation room. You'll need at least two computers: one for the moderator and another to mirror the participants' webcam and desktop video. Set up the following on the moderator computer:

1. Software to voice chat with your participant (Skype, or TeamSpeak if you're going to be testing more than one participant at the same time)

2. A microphone headset for communicating with the user

3. Any instant messaging client you can use to communicate with moderators and observers

4. A word processor to take notes with and display your facilitator guide

On the other observation room computer, you'll want to install VNC, a remote desktop control tool you can use to mirror the participant's webcam output, and a projector, if you want to display it on the wall.

Finally, set up a monitor that will connect to the participants' computer so you can mirror their computer screen.

Now get the participants' station ready. Because this is "simulated native environment" research, there's an aesthetic component to setting up the participants' station. You don't need to go as far as to make your testing area look like a "real living room," whatever that is, but neither should it look too much like an office or a lab. Soft floor lighting and a comfortable chair and desk can go a long way; aim for informal and low-key.

The participants' computer is going to need a lot of careful preparation. You'll want to set it up with the following:

1. Skype (or TeamSpeak, depending on what the moderator is using)

2. A microphone headset for communicating with the moderator

3. VNC, to be able to connect with the observation room computer

4. A Digital Video Recorder (DVR) unit, to record the computer's audio/video output (we used the Archos TV+)

5. The interface to be tested

6. A VGA cable long enough to reach the observation room

You'll also want to make sure there's nothing else on the computer desktop that might disrupt the participants' experience, especially anything that reminds participants that they're in a research study. Shortcuts to the interface you're testing should be fine, but otherwise try to stick to the default computer appearance.

Finally, there's the participants' computer. In this day and age, chances are that your participants' computer won't be fast enough to handle running a recording webcam, a screen recorder, VNC screen sharing, voice chat, and the test interface all at the same time without negatively affecting computer performance, so using a second computer to delegate the computing tasks is an option. On this second computer, you can run the webcam and any other testing materials, such as the surveys we used in our Spore study. You can also run the voice chat program on this computer, if it's not important to record the sound on your interface. If your webcam has a built-in microphone, enable it. It's good to have a redundant audio stream in case the TeamSpeak/Skype recording fails. Table 9.1 breaks down what equipment is necessary at each location in a simulated native environment.

TABLE 9.1 SIMULATED NATIVE ENVIRONMENT EQUIPMENT
SETUP AT A GLANCE

Location	Zone	Equipment/Software
Observation room	Moderator computer	TeamSpeak
		Microphone headset
		Any word processor for notes
		Facilitator guide
		IM to chat with moderators and observers
	Observation computer	VNC viewer
		Projector
	Room	Headphones for observers
		One monitor per participant
Each participant station	Participant computer	TeamSpeak
		VNC server
		Microphone headset
		DVR unit (we used Archos TV+)
		Extra VGA monitor out, with monitor cable long enough to reach observation room (if not using VNC to share desktop)
		Interface to be tested
	Station computer	Webcam
		Survey, full-screen mode (optional)
		Touchscreen monitor (optional)
		VNC server
		Screen recording software

Pros and Cons of Simulated Native Environments

Simulated native environments address a lot of the things that personally bug us about standard lab testing. The major benefits to this approach are the physical absence of moderators and other participants, which eases the Hawthorne effect and groupthink. In this study, the users mostly weren't prompted to respond to focus questions or perform specific tasks, but instead just voiced their thoughts aloud unprompted, giving us insight into the things they noticed most about the game, instead of what we just assumed were the most important elements.

It's still worth noting the shortcomings of this approach compared to true remote methods, though. You're still testing users on a prepared computer, rather than their own home computer; it's not really feasible to *precisely* replicate the experience of being at home or in an office because homes and offices obviously vary. The equipment overhead for this kind of testing is also a lot greater than for typical remote testing. At the very least, you'll need a station for each user, equipped with computer, webcam, Internet access, and microphone headset. And, of course, you're still restricted by all the physical and geographic predicaments of lab research, which means you can have no-shows, people may show up late, and so on.

Actually We Messed Up a Little

Over the year-long course of the *Spore* research project, one incident proved to us just how important it was to preserve the self-directed task structure of our research. Since *Spore* is divided into five discrete "phases," we initially believed it was important to structure the sessions in a way that gave players a chance to try each phase for a predetermined amount of time, in a set order. Partway through the second session, we started having doubts. Even though we weren't telling users what to do within each phase, what if our rigid structure and sequencing were affecting the gamers' engagement and involvement with the game?

To minimize this situation, we made a significant change to the study design between sessions. Instead of telling users to stop at predetermined intervals and proceed to the next phase of the game, we threw out the timing altogether and allowed users to play any part of the game they wanted, for as long as they wanted, in whatever order they wanted, and the only stipulation was that they should try each phase at least once. Each session lasted six hours split over two nights, so there was more than enough time to cover all five phases, even without prompting users to do so.

Sure enough, we saw major differences in player feedback, in both the players' comments and their survey responses. Unfortunately, for legal reasons we can't provide much in the way of specific findings, but we can say that the ratings for certain phases consistently improved over previous sessions, and a few of the lukewarm comments players had made about certain aspects of the game seemed to stem from the limiting research format, rather than the game itself.

It became clear that when conducting research on games, sticking to the actual realities of natural gameplay was more important than following precisely structured research formatting. You have to loosen up a little bit; video games are, after all, interactive and fun. It makes no more sense to rigidify a gameplay experience than it does to add sound effects and animated 3D graphics to a spreadsheet application.

Bringing Game Research Home

There are still many ways to go with the native environment approach. Even with all our efforts to keep the process as natural and unobtrusive as possible, there are still opportunities to bring the experience even closer to players' typical behaviors. Obviously, we'd love to do game research remotely, allowing participants to play right at home, on their own systems, without even getting up.

Consider, if you will, what a hypothetical in-home game research session might be like: a player logs in to an online video game service like XBox Live and is greeted with a pop-up inviting him/her to participate in a one-hour user research study to earn 8,000 XBox Live points. (The pop-up is configured to appear only to players whose accounts are of legal age, to avoid issues of consent with minors.) The player agrees and is automatically connected by voice chat to a research moderator, who is standing by. While the game is being securely delivered and installed to the player's XBox, the moderator introduces the player to the study and gets consent to record the session. Once the game is finished installing, the player tests the game for an hour, giving his/her think-aloud feedback the entire time, while the moderator takes notes and records the session. At the end of the session, the game is automatically and completely uninstalled from the player's XBox, and the XBox Live points are instantly awarded to the player's account.

Remote research on PC games is already feasible, and for games with built-in chat and replay functionality, the logistics should be much easier to manage. Console game research, on the other hand, will likely require a substantial investment by console developers to make this possible, and handheld consoles present even more challenges. Lots of basic infrastructure and social advances would need to happen before our hypothetical scenario becomes viable:

- Widespread game console broadband connectivity

- Participant access to voice chat equipment

- An online live recruiting mechanism for games, preferably integrated into an online gaming framework

- Secure digital delivery of prototype or test build content (may be easy with upcoming server-side processing game services like OnLive)

- Gameplay screen sharing or mirroring

As game consoles become more and more like fully featured PCs and content delivery platforms, we expect that most of these barriers will fall in good time. We also believe that allowing players to give feedback at home, the most natural environment for gameplay, would yield the most natural feedback, bringing game evaluation and testing further into the domain of good UX research.

Portable Testing (Cars)

One of the major limitations of current remote testing methods is that you're restricted to testing computer software interfaces. Screen sharing won't give you the whole picture if you're trying to test a physical interface, especially a mobile one like a cell phone. On the bright side, technology and Internet access get more portable and mobile every day, and researchers are starting to figure out ways to embed technology into the physical world to make research more flexible and natural. There's been plenty of mobile device research in the past, but it's been largely focused on quantitative data gathering, random "experience sampling," ethnographic interviews, heavily structured lab research, or usability-focused studies. Comparatively little has been done to directly observe and document rich natural user behavior with mobile or location-dependent interfaces, such as in-car GPS systems, mobile devices, and cell phones.

In 2008, we conducted an ethnographic study on behalf of a major auto manufacturer whose R&D team wanted to understand two things: first, to observe how people use different technologies in the car over the course of everyday driving—cell phones, in-car radios, iPods, GPS systems, and anything else that drivers might bring in; and second, to understand what needs these usages fulfill. The R&D team wanted to be involved in the sessions as they were ongoing (as they would in a normal remote study), but we obviously couldn't have the whole team pile into the car with the passenger and moderator. The solution was surprisingly straightforward (see Figure 9.6).

FIGURE 9.6
The moderator rides
along in the back seat,
observing and filming
the driver. The footage
is streamed live to
the Web.

Method

The idea was to conduct a standard ride-along automobile ethnography, with a moderator interviewing a participant *in situ*, but also to use an EVDO Internet connection (go-anywhere wireless Internet access), a laptop, and a webcam to broadcast the sessions and communicate with the clients. (We also had a cameraman accompany us to document the experience in high-definition video, but that was mostly for our own benefit.) To broadcast, we signed up for an account on Stickam, a live Web video streaming service (others include Justin.tv, UStream.tv, and Qik).

We recruited our users in advance from the Los Angeles and San Francisco Bay areas through a third-party recruiting agency, selecting them on the basis of gender (equal balance), age (25–55), and level of personal technology usage (high vs. low).

The setup was very bare bones, but here's a checklist of our equipment, all of which is handy to bring along:

1. Sierra Wireless 57E EVDO ExpressCard

2. Logitech webcam

3. Web video streaming account (we used Stickam)

4. A fully charged laptop with extra batteries

5. A point-and-shoot digital camera to take quick snapshots for later reference

Our methods weren't actually all that different from a normal moderated study: we had users think aloud as they went about their tasks (while still following all road safety precautions, of course) and also explain to us how and why they used devices in the ways that they did (see Figure 9.7). For the purposes of our study, we found it helpful to focus on a few behavioral areas that were often very revealing of participants' usages of technology:

- At what points did users engage with their devices?

- How did device usage affect other device interactions?

- How did they combine or modify devices to achieve a particular effect?

- What functions did they use the most, and which were they prevented from using?

- Who else rides in the car (kids, pets), and how did their presence influence their capacity and inclination to use personal technologies?

FIGURE 9.7
A driver multitasks
while stopped in traffic.

Challenges and Logistics

Researching users in a moving environment is different from ethnographies you'd perform in a workspace or home, in the sense that some elements of the users' "surroundings" are constantly changing, while others remain stable. Even though we had to keep our attention mostly on what the users were doing with their technology and their driving behavior, often we encountered situations when what was happening outside the car affected the users' behavior. For example, one user who was fiddling with her iPod hit a sudden bump in the road, startling her and putting her attention back to driving.

Which reminds us—accidents do happen. We made sure to have participants read and sign two liability release waivers before the study, to cover our behinds in case the participants got into an accident. One form was an easy-to-understand, straightforward one-pager, and the other was a stronger and more exhaustive legal document. We didn't want participants blaming us for causing the accident by distracting them from their driving. (Accordingly, this is another good reason why the moderator should stay as unobtrusive as possible.)

The car environment allowed for reasonable laptop usage, but many mobile technologies would make it impractical to use a laptop while keeping up with users, especially when standing or walking. In those situations, you may want to concentrate on recording the session and transcribe it later, though this sort of transcription can make the analysis process a whole lot lengthier. If you can afford it, we recommend using an online transcription service such as CastingWords.

Another issue is wireless Web connection: depending on the locations, distances, and speeds traveled during the sessions, EVDO access may become unstable or spotty. The study should be designed with accessible coverage in mind, whether via wireless connection, EVDO, or even cell phone tethering (i.e., using your cell phone to connect your laptop to the Internet), which will probably be getting more popular and accessible in the coming years. If you'll be driving through a tunnel, getting on a subway, etc., expect connection outages.

As always, keeping the task as natural as possible is important for understanding how users will interact with a device naturally. As with simulated environment testing, we recommend downplaying the moderator's physical presence or keeping the moderator out of the users' field of vision entirely; our moderator rode in the back seat. Since the study was scheduled, we asked the users to choose a time during which they'd be doing something completely typical in their car, whether going to the grocery store, picking up kids, commuting, and so on. This approach helped us reduce the amount of moderator involvement in dictating the tasks and allowed us to focus on the natural tasks. In our car study, when users decided midsession to change their mind about where they wanted to go or what they wanted to do in their car, we made sure the moderator intervened only if that would have completely deviated from the goals of the study, which didn't happen. Otherwise, you should just accept it as an example of good ol' real-world behavior.

Getting Better All the Time

In our study, the quality of live streaming video to remote observers was limited by bandwidth issues, both because of the coverage problems we just described and because EVDO wasn't really implemented with hardcore high-definition video streaming in mind (see Figure 9.8). We'll all probably have to wait a few more years before the reliability and quality of mobile livecasts becomes a nonissue.

FIGURE 9.8
Side-by-side EVDO and high-definition footage.

And then there's moderator presence, which is a liability in many ways we've already discussed. Not to get too Big Brother-y, but the development of more powerful and portable monitoring tools—light, wearable livecasting equipment—can eliminate the moderator's presence altogether, allowing users to communicate with the moderator remotely through earpieces or videophones or microphone arrays or whatever other device best suits the study. This will not only resolve the issue of moderator disruption and intrusion, but also opens up new domains for mobile user observations, including longer-term ethnographies (during which the researchers can "pop in" at specified times to eavesdrop on natural behavior), as well as studies of situations in which physical accompaniment is difficult or impossible (e.g., on a motorcycle, in a crowd, etc). It's all a matter of time until talking to users anywhere, doing practically anything, will be as simple as having them wear a little clip-on camera as they go about living their lives.

Staying Current

It's fitting that the practice of observing how people use new technology requires the researchers themselves to use new technology to its fullest; that's what remote research methods are about, really. To that end, you (and all user researchers) should always be on the lookout for new tools, services, devices, and trends in mobile technology that you can exploit to help capture user behaviors that are as close as possible to real life.

Chapter Summary

- You can use various aspects of remote research and new technology to adapt to novel testing challenges.

- Secure interfaces can be tested with a method called "reverse screensharing": allowing your participants to view and take control of one of your computers, so that the code remains secure on your end.

- Mobile interface research will be huge and will offer new opportunities for Time-Aware Research; there will be many new considerations for mobile testing.

- Even if you conduct research in a lab, you can use webcams and screen sharing to remove the moderator's physical presence and conduct multiple one-on-one sessions simultaneously.

- To test interfaces in moving environments like cars, you can use a wireless EVDO connection and webcasting services to record and stream sessions, and also communicate with clients.

CHAPTER 10

The Challenges of Remote Testing

Y ou've seen how remote research deals with the problems of traditional in-person research (geographical distance, recruiting, task validity, etc.), but it raises plenty of its own problems, too. We'd like to wind down this discussion with a review of the biggest challenges of adopting remote research methods: the doubts, concerns, and pains in the neck that seem to come up in study after study, even for seasoned practitioners.

Legitimacy

Remote research is still in its adolescence, and skeptical prospective clients often ask us, "Who else does remote research? If it's so great, why haven't *I* heard of it?"

As we mentioned at the beginning, lab research has run the show for a long time mostly because that's the way things have always been done. In spite of this, plenty of big-name corporations have happily taken the plunge with remote research, including (just from personal experience) Sony, Autodesk, Greenpeace, AAA, HP, Genentech, Citibank, Wikipedia, UCSF Medical Center, the *Washington Post*, Esurance, Princess Cruises, Hallmark, Oracle, Blue Shield of California, Dolby, and the California Health Care Foundation, to name but a scant few. Automated tool sites boast Sony Ericsson, Motorola, YouTube, REI, eBay, Cisco, Bath & Body Works, Orbitz, Hyundai, and Continental Airlines as customers.

If you're still not sure about it, we recommend looking through an exhaustively documented study (complete with full-session videos and highlight clips) of first-time Wikipedia editors that we conducted for the Wikipedia Usability Initiative project (http://usability.wikimedia.org/wiki/ Usability_and_Experience_Study). It includes both lab and remote sessions with identical goals, so it's a good comparative case study. If you have any reservations, you should watch the sessions and decide for yourself.

Not Seeing the Users' Faces

We have always been confident that seeing a user's face isn't necessary for gleaning insight out of a user study, but clients and stakeholders and some UX researchers can get very persnickety about this issue. If a person isn't physically present and being videotaped or sitting behind glass, they wonder, "How can you really research them?" Or "How can you develop empathy for someone you can't see?"

Our firm belief is that onscreen user behavior and think-aloud comments, and not users' facial expressions, provide the real value of the study because you want to learn what the users do and are able to do on the site, not how they feel about it. Even if we concede that participants' emotional responses can bring in valuable insights about how they use a site, you'd be surprised at how much feeling can be just as effectively inferred from vocal tone, sighs, pauses, inflections, and interjections ("Ah-ha!" "Oh, *man!*"), not to mention the content of what they're saying. Most people are veteran telephone users and have learned by now how to express themselves vocally.

Maybe in a few years, video chat will be commonplace, and not seeing the users' faces probably won't even be an issue anymore. For now, however, rest assured that not seeing the user's face just isn't that big a deal.

Technology Failures

Moderated remote research uses lots of separate technological components, any of which can malfunction for many reasons: a computer with multiple programs running on it, a microphone headset, an Internet connection, Web recruiting tools, third-party screen sharing solutions, recording software, two phone lines, IM clients, and so on. Then there are all the things that can go wrong with the *users'* computer and phone setup. Users can be on a wireless connection, an unstable wired connection, or a cell phone; international phone lines can be muddy; their computers might not be able to install or run screen sharing.

One or two things going awry amounts to annoying delays: interruptions to an ongoing study, glitches in the recordings, difficulty hearing users, and so on. At its worst, having two or three or all these things fail can stop a study cold until the problems are resolved.

UX researchers aren't necessarily tech experts, so if you want to stave off these problems, the best thing to do is test everything at least a day prior to the start of the study, referring to a checklist. Table 10.1 is a starter checklist for you. Modify it to suit the tools you use to conduct your research.

TABLE 10.1 TROUBLESHOOTING CHECKLIST

Problem	What to Do
Screen sharing is interrupted/ malfunctions	Check to see whether your Internet connection is stable. Check to see whether your user's Internet connection is stable; if possible, have him/her switch to a wired connection. If it's still not working, try a different screen sharing tool.
Recordings come out corrupted/ glitchy/ truncated	Test the recording tool. If test recordings don't work, check recorder settings to ensure that they are recording to the correct format and quality. If test recordings work fine, most likely the computer was running too many processes during the recording. Close down unnecessary programs to fix this problem; if it persists, you may need to upgrade your computer with more RAM. Also check whether you have sufficient hard drive space to store the recordings. For corrupted files, use a video editing program or converter to attempt to convert the file to a different format. For certain file formats, there are also utilities that are able to fix minor problems.
Phone connection malfunctions	Check your phone connection. Use an alternate phone line, if one is available. Ask users if they're using a cell phone. Ask if there is an alternate line to call. Ask users whether you can call back on a different line, at another time if necessary.
Microphone headset/ sound input malfunction	Check whether the headset is muted. Check the mic input volume in system settings. If you're using a VoIP service like Skype, check the software settings to see if it's not muted.
Internet connection seems choppy or breaks	The problem could be either your connection or the user's. If it's yours, postpone the study and switch to any alternate Internet connections you may have in your office. As a last resort, call your Internet service provider and see if the service has gone down. If it's the user's connection, ask the user if he/she is on a wireless connection; if so, ask if he/she is able to switch to a wired connection. If that doesn't work, attempt to reschedule the study to a time when the user will be at a different computer.

TABLE 10.1 TROUBLESHOOTING CHECKLIST
(CONTINUED)

User's firewall does not permit the screen sharing tool to function	Switch to an alternate, preferably browser-based screen sharing solution. If no available solutions, attempt to reschedule the study to a time when the user will be at a different computer.
Recordings have no sound	Check the system sound input volume and settings (make sure it's not muted) and recording software settings.

Regardless of what happens, *stay calm*. The absolute best way to handle technical problems is to set everyone's expectations ahead of time (yours, your team's, and those of anyone who's observing) that there's always a chance issues will come up, and that it's a normal part of the process. Make sure observers have their laptops or some poetry to read so they don't sit around idly when a user's cell phone dies.

In spite of your planning, it's always stressful when you have observers watching you, a live participant waiting on the other line, and a stupid technological problem interrupts everything, even though you're *positive* you tested it *like a million times*. Take a few seconds to step back and put it in perspective: life goes on. A hard-breathing, hyperthyroidal moderator will spoil a session even if all the technology starts working again.

Not as Inexpensive as You'd Think

Remote research is often represented as a discount method, a way of shaving costs, and people are often surprised to find that the cost of a remote moderated study is usually comparable to its in-person equivalent. Remote research can help save on travel, recruiting, and lab rental costs, but where moderator time, participant incentives, and scheduling are concerned, nothing is much different. Most of the expense of a research project is the research—having a trained researcher take the time to observe and analyze users' behaviors carefully and then synthesize the findings into smart and meaningful recommendations. Don't let the stakeholders of the study fall under the impression that the primary motive behind a remote study is the cost savings: the real benefit, again, is its ability to conduct time-aware research.

Organizational Challenges of Web Recruiting

Most Web recruiting tools require you to place a few lines of external code in the Web site's source code. If you have a personal Web site or work for a small, scrappy start-up and have direct access to the code, this task shouldn't be difficult. If, on the other hand, you're dealing with a huge company with a complex content management system, you may have to prepare for red tape. You'll have to cooperate with the IT operations guys and higher-up managers who have the final say as to what goes on the Web site. Be sure you have answers to the following questions:

- What are the changes we need to make to the code?

- What does the code do? Is it secure?

- What pages does the code need to go on? Will it work with our Content Management System?

- Which pages will the screener appear on?

- How long will the recruiting code be active?

- What will the screener look like to visitors?

- How many people will see it?

- How can the managers/IT people shut it off or disable it on their end?

- Will the look and feel of the screener match the Web site's look and feel?

The answers to all these questions depend on the tool you're using to recruit. Come prepared with the answers to these questions before meeting with your IT people/managers to prevent delays and confusion in getting the screener up.

Getting the Right Recruits

Taking matters into your own hands with live recruiting on the Web is often cheaper, faster, and more dependable for remote research, but it means that you'll have to bear more responsibility for recruiting your participants properly. For any number of reasons, getting enough recruits to conduct steady back-to-back sessions may not be easy for you. See Table 10.2 for reasons why.

TABLE 10.2 DEALING WITH SLOW RECRUITING

Problem	What to Do
Your Web site's traffic volume isn't high enough to bring in six qualified recruits an hour	Increase the screener display rate if it's below 100%. Place the screener on multiple pages or a higher-level page in the IA. Schedule qualified recruits in advance to supplement the users you're able to obtain. Increase the incentive, but not too high (or else you attract more fakers). Lengthen the duration of the study (with healthy traffic, it's possible to do about six users in a work day).
Your recruiting criteria are too strict	If you're filtering your results, disable the filter to see if any of the filtered recruits are acceptable participants. Ask stakeholders if any recruiting criteria are negotiable and relax the lowest-priority ones. Increase the incentive.
The wording or length of your recruiting screener turns people off	Revise the wording to feel less like a deal or an offer. Omit needless words and questions. Be specific about the incentive.
Fakers are filling out your recruiting form	Review the "Why did you come to this site?" responses to determine whether the fakers were referred by a deals/bargains site. Add sneaky questions to the screener to trick fakers into tipping their hand. Add open-ended questions that can be answered plausibly only by your legitimate recruiting audience.

Natural User Behavior

Moderated remote research is great for watching users perform natural, self-directed tasks on their own initiative, but that kind of behavior isn't a given. Some users who participate in a study have preconceived notions about what's expected of them during a study. Either they'll tell you what you want to hear, or they'll be too critical. Some will ask: "So, what do you want me to do?" At every turn, you should encourage users to do what they would naturally do, only adding that you may have to move things along to keep to the time limit of the session. (This is a polite way of warning them that you might cut them off if they start meandering.)

When users get absorbed in their tasks, they may stop thinking aloud. That's not necessarily a bad thing, depending on how clear their motivations are. Usually, you can keep users talkative with a few encouraging reminders to "keep saying whatever's going through your head about what you're doing." Naturally quiet or shy users might need more explicit prompts, with extra acknowledgment of how awkward it is to think aloud: "I know it's kind of odd to talk constantly while you're browsing, but everything you have to say is really useful to us, so don't hold back."

Then again, sometimes it's not the users who have problems with natural behavior, but the stakeholders. For an outside observer who's accustomed to heavily scripted and controlled lab testing approaches ("Now do Task A... Now do Task B..."), it can be jarring to watch participants use the interface the way they normally would. Observing natural behavior often means allowing users to go off on digressions or to allow for long silences while users try to figure something out or to perform tasks that don't appear to relate to the scripted tasks.

You need to set your stakeholders' expectations. What may seem aimless and chaotic is actually rich, properly contextualized interaction that they should pay close attention to. Put it this way: when you go to a Web site, do you close down all your other applications and browser tabs, turn off your cell phone, stick to one focused task, and tell the kids and dog to be quiet? And even if you do, is anyone *ordering* you to do those things? You need to assure stakeholders that regardless of whatever unanticipated tasks the users perform, the moderator will see to it that the users also perform the core, necessary tasks.

But there are some cases in which users really *are* too distracted to be paying any attention to what they claim to be doing. If they're simply veering off-track, you may either need to reschedule the session when it's less hectic for these users or dismiss them. That decision is up to the discretion of the moderator, but it's usually pretty obvious. Whether users listen and respond to what the moderator says is often a good indicator.

Multitasking

It's tough to appreciate, without doing a few sessions, how much stuff you have to keep your eye on while moderating a remote session: your conversation with the user, the user's onscreen behavior, observer questions and comments via IM, your notes, the time remaining in the session, your place in the facilitator guide, and occasionally the status of the recording tool. You also have to exude an aura of serenity; you can't even *sound* as though you're trying.

The main thing is practice, practice, practice. Find willing volunteers to participate in dry runs. Watch and learn from recorded past sessions—like our sessions from the Wikipedia Usability Initiative (http://usability.wikimedia. org/wiki/Usability_and_Experience_Study#Remote_Testing), for example.

Security and Confidentiality

Finally, there are the challenges of testing interfaces that need to be presented to users securely. These interfaces can't be installed on users' computers or placed live on the site, usually because they're prototypes that aren't ready for public exposure. Password-protected access to the site is the most preferable option, but in cases in which no files can be moved to users' computers, you should use the reverse screen sharing techniques described in Chapter 9, making sure that the Internet connection is fast enough to support a natural interaction.

Persistent Negativity

Sometimes, for no particular reason, you'll have stakeholders or team members who think remote research is a horrible and stupid idea. This opinion doesn't make them bad people. Even after a great study, there will sometimes be criticisms of some of the methods and details. The reason is largely that most people aren't familiar with remote research yet and don't know what a successful session looks like. They'll get freaked out about the moderator not assigning specific tasks, about having to wait 20 minutes to find a qualified user to live recruit, about the lack of active listening, or about any of the other things that are supposed to happen. And then there are die-hard skeptics, who won't like what they see no matter what.

The best remedy for dealing with these situations is to deliver amazingly successful findings, which exceed the usual expectations of incremental usability fixes. Of course, doing that is not easy, but in spite of anyone's doubts about the process, if you think hard about your users' behavior in context of their real lives and then come up with insights that double the conversion rate or dramatically increase the ease of use of your interface, the naysayers will be turned.

Chapter Summary

- Although remote research makes many things easier, it also introduces its own unique challenges for the researcher.

- Many people are still skeptical about remote research because it's new. Some people believe you can't get good results without seeing users' faces. (You can.) And some people are just plain resistant to the idea from the beginning. Smart effective findings will change their minds.

- Since remote methods use lots of technology, there's a higher incidence of tech failure. Be prepared for the most common scenarios.

- One misconception is that remote research is significantly cheaper than in-person testing. While it's true that you can save some costs, the overall cost is not drastically lower than in-person studies.

- In most medium-to-large organizations, you'll need to get different parts of the organization involved if you're going to use a screener to do live recruiting.

- It takes effort, patience, and experience to get the right recruits for your study, but as long as you have enough Web traffic, there are always things you can do to help things along.

- It's crucial to get people to behave naturally if you want good feedback. You have to pay attention to your phone mannerisms to make that happen. Reassure the study observers that going off-script and allowing silences is necessary to encourage natural behavior.

- You need to multitask heavily to be an effective remote moderator. Practice a lot and watch old session videos to improve your techniques.

- Confidentiality must be maintained for both you and your participant; take security precautions and use discretion in your language.

Don't Waste Your Life Doing Pointless Research

L ots of companies have come to accept user research as part of the iterative product development cycle—just another step in the refining process, preliminary Quality Assurance. Actually, what they've really embraced is usability research, which is a subdiscipline of the broader field of user research. Whether it's conducted in a lab or remotely, usability research is concerned with making things easy to use, findable, error-proof, and intuitive.

There's nothing wrong with usability, but the truth is, you don't come up with an iPod just by making a Walkman really easy to use. Mucking around with button placement and form alignment won't lead to the smash hits we all want to work on when developing interfaces. As Todd Wilkens wrote in an Adaptive Path blog post: "Usability is too low level, too focused on minutia. It can't compel people to be interested in interacting with your product or service. It can't make you compelling or really differentiate you from other organizations.... There's only so far you can get by streamlining the shopping cart on your Web site."

As a researcher, you are responsible not only for making practical usability findings, but also for keeping an eye out for the unexpected user behaviors that reveal deep design insights, as well as what users are really after. Be prepared to question the purpose of the interface you're seeking to improve. Is it more worthwhile to make your interface as easy as possible to use or to design something completely new that will potentially eliminate all those problems and more? That's a perfect question for a user research study, and one that often goes unasked.

The unique qualities of remote research can help you combat the tendency toward task-based usability nitpicking:

- **Time-aware research.** Live recruiting allows you to intercept and contact participants instantly, which means you're able to talk to users at the exact moment in their real lives that they're actually doing the tasks you're interested in watching. You don't have to order them to perform tasks they're not motivated to perform on an arbitrary schedule, thus making their behavior more representative of natural usage. That's the most groundbreaking difference between traditional and remote research: the promise of being able to leave behind artificial lab environments, time structure, and artificial tasks. Instead, you can focus on observing real-life interactions with technology as they happen.

- **Native environment.** Remote research is a practical way to observe users in an authentic context. Since users are at their own computers, they're more comfortable and have access to all the peripheral tools and artifacts they'd normally use to help them with their tasks. This gives you a wider, richer view of how users are really using your interface.

- **Remote behavioral observation.** Collecting simple, opinion-based user feedback over the Web has always been easy, but the introduction of screen sharing and automated Web tracking technologies makes it possible to see exactly how users are *behaving*. That makes it possible to gather evidence about users' real needs, problems, and workarounds, without having to conduct an in-person ethnography.

Those qualities make remote research methods great for conducting research that inspires innovation. It's not the only way to get inspired, however, and when new kinds of interfaces arise, consider what you can do to improve the methods yourself. We came up with the strategies in this book over almost a decade of refinement, but we're still developing strategies to accommodate all the new tools, interfaces, and research challenges that are constantly being introduced. In the same way that over-attachment to lab research prevents some people from seeing the value in remote methods, you shouldn't do remote research if another (potentially undiscovered) approach will give you a more truthful perspective on your users' behavior. Instead of sticking to a method because you're comfortable with it, find the method that will get the most accurate results.

Don't insist on a method.

Insist on doing things right.

Index

A

access, as Safe Harbor principle, 87
accidents, portable testing and, 233
acknowledgments of observers'
    contributions, 112
Acorn Marketing and Research
    Consultants, 84
active listening, 117
Adaptive Path, 8
Adium, 104
Adobe Acrobat Connect Pro, 184
Adobe Captivate, 201
Adobe Connect, 30, 33, 35, 183–185
    observer instructions on, 45
    session window, 31
    subscription plan, 34
    video conferencing, 37
Adobe Premiere, 105, 162
Albert, Bill, 127
Amazon.com gift certificates, 64
annotation tools, 150
Apogee Group Recruitment, 84
Apple, research, 177
area codes, checking for recurring, 65
asynchronous research, 122. *See also*
    automated research
automated analysis, 163–168
    task elicitation and
        analytics, 163–166
automated research,
    annotation tools, 150
    basics, 122
    different kinds, 127–128
    input tracking, 150
    vs. moderated research, 123
    multivariate testing, 149
    online diaries and ethnographies,
        149
    presenting findings, 172
    qualitative and quantitative
        methods, 132

recruiting for, 124
remote card sorting, 142–147
structure of study, 122–124
task elicitation, 129–142
Web analytics, 149
automated tools and services, 202–212
AVI files, 36, 197
Axure, 16

B

back button usage, 165
backup for screen sharing tools, 192
BaseCamp, 103
Beaver, Brian, 8
behaviors, vs. opinions, 5
*Beyond the Usability Lab: Conducting
    Large-Scale User Experience Studies*
    (Albert, Tullis, Tedesco), 127
blue jeans debacle, 81
bookmarks, 159
bored participant, 113
Brighton University, usability lab, 2
Budd, Andy, 21

C

CAMREC format, 197
CamStudio, 198–199
Card Sort Cluster Analysis Tool, 211
card sorting, 142–147
    analysis, 166
    moderated or automated, 145
    presenting findings, 173
    tools for, 209–211
cars, research on technology use in,
    230–235
CastingWords, 105, 233
Chalkmark, 204
charity, donation as recipient
    incentive, 64
chatting, 34
    GoToMeeting and, 191

"Check all that apply" statements, 133
children. *See* minors
Chisnell, Dana, Professional Recruiter's
  advice, 68
choice, in Safe Harbor principles, 86
Cisco WebEx, 194
clarity, in consent statement, 81
Clearleft, 21
Click Density, 202
ClickHeat, 202
ClickTale, 203
clickwrap agreement, 76, 89
Clixpy, 204
closed-ended questions, 132
  in automated research, sample, 137
closed sorts, 143
collaboration for note taking, 103
collectivism, as dimension of
  culture, 106
communication, high- and low-context,
  106
competitive analysis, 135
computer games. *See* games
computers
  at corporate institutions, use
    restriction, 160
  for moderated research, 29
  note taking on, 103
"concept testing", 150
conference call in GoToMeeting
  two-line desk phone for, 190
confidentiality, 245
Connect. *See* Adobe Connect
"consent to contact" question, 60
consent agreement, 76
  for recording, 82–84
  invalid, 77
  sample, 80
consent from participants, 74. *See also*
  international consent
  basics, 76–82
  for minors, 88–89
  and trust, 83

context-dependent interfaces, 15
Cooper-Symons Associates, 84
costs
  of remote research, 241
  of screen sharing software, 34
  savings from remote research,
    10, 14
  for testing labs, 21
craigslist, 126
CrazyEgg, 204
culture, dimensions of, 106
cultures, testing across, 106, 112, 115

D

data analysis of findings, 127
data integrity, as Safe Harbor principle,
  87
debriefing sessions, 111
The Designful Company (Neumeier),
  178
design, recruiting screener, 59–63
desktop wallpapers, 160
DHTML layer, 54
DHTML recruiting screener, 57
Digsby, 104
downtime between sessions, observers'
  expectations on, 44
drop-down menus, in screener, 61
drop-off rate for task, 165
due diligence, 85
"dummy" information, user entry of, 16

E

eCamm's Call Recorder for Mac, 36
email
  for automated research, 125
email domains, checking for
  recurring, 65
email for recruiting, 50
"emerging themes", 101
enforcement, of Safe Harbor principles,
  87

Player Experience of Need Satisfaction
(PENS), 223
plug-in usage, 161
PocketMeeting, 220
PollDaddy, 207
pop-up blockers, 54
pop-ups, for live recruiting, 52
portable testing, 230–235
  challenges and logistics, 233–234
  method, 231–232
Power Distance, as dimension of
  culture, 106
privacy
  software bugs and, 81
  spying and, 99–100
privacy policy
  for participants' information, 75
  for Web site, 75
professional survey-takers, 65
"progressive disclosure of
  information", 99
prototype software
  remote research on, 16
  user's remote access to, 216
purpose of research, explaining, 99

Q

Qik, 231
qualitative data in reports, 173
qualitative feedback, 165
qualitative methods of automated
  research, 132
QualVu, 149, 195
quantative methods of automated
  research, 132
quantitative data in reports, 173
questions
  fielding from newbies, 175
  in automated research, sample, 137
  in automatic research, phrasing, 131
  in recruiting screener, 60–62
Quicktime Pro, 201

quiet environment
  for moderated research, 30
quiet participants, 113, 115
  encouraging to talk, 244

R

rapport with users, 94
RealPlayer, 162
recording, 35
  avoiding with minors, 91
  consent agreement for, 82–84
  consent for, 40
  hard drive space for, 198
  troubleshooting problems, 240
  with UserVue, 186
recording software, 196–202
  CamStudio, 198–199
  Camtasia Studio/Camtasia for
    Mac, 196–198
recruiting, 9
  adjusting screener to increase
    numbers, 67–70
  for automated research, 124–127
  basics, 48
  dealing with slow, 243
  of first-timers, 51
  for card sort, 146
  for international participants, 84
  for lab research, 19
  live, 48–50, 243
  methods compared, 50
  organizational challenges, 242
recruiting agencies, 50
  for contacting minors, 89
  time requirements, 14
recruiting funnel, 56
recruiting screener
  implementing, 52
  logistics of implementing, 54–55
  placement in Web site, 68
  requiring completed forms before
    submission, 61

ACKNOWLEDGMENTS

Louis Rosenfeld of Rosenfeld Media went to bat for our book before it was even a book, and surely allowed this project to become what it is: the greatest work of literature in any language.

Marta Justak, our wonderful main editor, initiated us into the terrifying and desolate world of technical book writing, and also forgave our San Francisco liberal attitudes *vis-à-vis* deadlines and template conformity. Our technical editors, Chauncey Wilson (Chapters 1–5, 7–9) and Donna Tedesco (Chapters 6–8), helped us dig up references, challenged sloppy thinking, put our thoughts in train, and kept it very real.

Many thanks to all our contributors: Peter Merholz, Andy Budd, Brian Beaver, Julia Houck-Whitaker, Dana Chisnell, Emilie Gould, and Carol Farnsworth. They wrote the words we were simply too afraid to write. Thanks also to Elizabeth Bacon from Devise for providing great supplementary info for Chapter 8 and to Danny Hope for providing photos of the Brighton University Usability Lab.

The early feedback of our preview editors—Lori Baker, Kyle Soucy, and Caleb Brown[md]helped us to look smarter than we are. Also thanks to our amazing Bolt | Peters team of remote pioneers—Cyd Harrell, Frances James, Kate Nartker, Julia Houck-Whitaker, Mike Towber, Julian Wixson, Brian Enright, and Alana Pechon[md]for battling out remote testing over the years and for their practical UX research insights, extracted by the authors with many a harassing early-AM phone call. Now the advisors: many thanks to Donna Spencer and her book, *Card Sorting,* for the professional advice on (what else but) card sorting. Elizabeth Bacon offered lots of tool references for Chapter 8. Matt Thomson of Kronenberger | Burgoyne advised us on matters of Internet privacy, recording, and consent law. Susan Kornfield also provided much helpful guidance.

Part of Chapter 5 was adapted from an article Tony Tulathimutte and Cyd Harrell wrote for *UPA Magazine* (Vol. 7.3). Parts of Chapter 9 were adapted from our earlier writing, which appeared on the Web site Boxes and Arrows (www.boxesandarrows.com) and in a paper for IA Summit 2008 called "Portable Research."

ABOUT THE AUTHORS

Nate Bolt

Nate is an entrepreneur and artist who likes to research and design things as El Presidente of Bolt | Peters. He has overseen hundreds of remote user research studies for Sony, Oracle, HP, Greenpeace, Electronic Arts, and others. In 1999 he began experimenting with remote research with sticks and string, and in 2003, he masterminded Ethnio: the very first moderated remote research app, which is now being used around the world to recruit hundreds of thousands of live participants for research.

Nate gives talks about user experience research and design in both commercial and academic settings, ranging from a keynote on "the future of library user experience" for the Urban Libraries Council to workshops for the Usability Professionals Association. He worked with faculty at the University of California, San Diego, to create a degree titled "Digital Technology and Society," which focused on the social impact of technology. He also completed a year of communications studies at the Sorbonne in Paris, where he failed every class and was jailed briefly for playing drums in public without a license.

Tony Tulathimutte

After getting a master's degree in Symbolic Systems at Stanford, Tony Tulathimutte worked as a user researcher for NHN USA, and joined Bolt | Peters in 2007. In addition to working on remote user research projects for AAA, Hewlett-Packard, *Harvard Business Review*, ANSYS, Autodesk, and Princess Cruises, he was also the lead researcher on the player experience study for Electronic Arts' 2008 game, *Spore*.

Tony is also a fiction writer, and in 2008 received an O. Henry Award for a short story. He spends his evenings luxuriating in a glycerin bath with an aged Chianti and a copy of the *Collected Keats*, tossing back a head of rich auburn hair and laughing, always laughing, at the caprices of the Fates.